Oral History and the Environment

Oral History and the Environment

Global Perspectives on Climate, Connection, and Catastrophe

Edited by

STEPHEN M. SLOAN AND MARK CAVE

OXFORD
UNIVERSITY PRESS

OXFORD
UNIVERSITY PRESS

Oxford University Press is a department of the University of Oxford. It furthers the University's objective of excellence in research, scholarship, and education by publishing worldwide. Oxford is a registered trade mark of Oxford University Press in the UK and certain other countries.

Published in the United States of America by Oxford University Press
198 Madison Avenue, New York, NY 10016, United States of America.

Library of Congress Cataloging-in-Publication Data
Names: Sloan, Stephen M., editor. | Cave, Mark, 1964- editor.
Title: Oral history and the environment : global perspectives on climate, connection, and catastrophe / edited by Stephen M. Sloan and Mark Cave.
Description: New York : Oxford University Press, 2022. |
Series: Oxford oral history series | Includes bibliographical references and index.
Identifiers: LCCN 2022018773 (print) | LCCN 2022018774 (ebook) |
ISBN 9780190684969 (hardback) | ISBN 9780190684976 (paperback) |
ISBN 9780190684990 (epub) | ISBN 9780197668108
Subjects: LCSH: Environmental sciences—Biographical methods. |
Human ecology—Case studies. | Environmental degradation—Case studies. |
Environmental disasters—Case studies. | Environmental protection—Case studies. |
Oral history.
Classification: LCC GE45.B56 O73 2022 (print) | LCC GE45.B56 (ebook) |
DDC 304.2—dc23/eng/20220718
LC record available at https://lccn.loc.gov/2022018773
LC ebook record available at https://lccn.loc.gov/2022018774

DOI: 10.1093/oso/9780190684969.001.0001

For those who lost their lives to Covid-19, including Hazra Bi, Hamida Bi, and Jabbar Khan; and all those narrators who never had a chance to share their stories.

Contents

Acknowledgments

Working on this, my second book with Stephen Sloan, has been a wonderful adventure. I feel so fortunate to have had a chance to work with such a patient and thoughtful co-editor. This is also my second time working with Nancy Toff of OUP, who has guided Stephen and me from start to finish (not an easy task). I want to thank my wonderful wife Patricia Murphy, who has nurtured a menagerie of ducks, herons, possums, turtles, and many little birds that bring us joy from dawn to dusk. Thanks to my parents Gilbert and Kathleen, who passed away during our work on this volume. They raised my sister Terri Potter and me on a farm in Ohio, giving us many magical moments in the natural world. The Historic New Orleans Collection, my employer for over twenty-five years, has been dedicated to our oral history program and has always given me a chance to explore. And last, but not least, Monkey (a cat), who has been my constant companion for the past fourteen years. She has had a paw in the editing of every chapter of this book.

—Mark Cave

If you ever get the opportunity to collaborate with Mark Cave on a project, jump at the chance. It has been my great fortune to get to do it twice. After the success of *Listening on the Edge: Oral History in the Aftermath of Crisis*, I approached Mark about doing a project to highlight the exciting research that was being done on the environment around the globe using oral history. His expertise, standing in the field, and his connections through his service to the International Oral History Association were indispensable in producing the present volume. Early conversations with oral historian Jeff D. Corrigan were also important to encourage me to do this project. I'd also like to thank my intrepid colleagues at the Institute for Oral History at Baylor University—Adrienne Cain, Michelle Holland, Dianne Reyes, and Steven Sielaff. For this project, the guidance of Nancy Toff and the patience of each of our contributors has been greatly appreciated. The University Research Committee in the Vice Provost for Research's office at Baylor University helped support this volume, and the efforts of graduate students Abigail Higgins and Annie Himler were essential in compiling the bibliography. Most importantly, I'd like to thank my always supportive and encouraging family, Melissa, Jacob, and Riley. Sharing life with you is bliss.

—Stephen M. Sloan

Contributors

Deb Anderson is an academic and journalist based in Melbourne. Born in Australia's northern Wet Tropics, she spent more than a decade working with the press in Australia and overseas, principally for *The Age*, before joining Monash University as a lecturer in 2013. Her research explores the lived experience of extreme weather for ordinary Australians, and perceptions of climate change in an era of politicized knowledge on the environment. She is the author of *Endurance: Australian Stories of Drought*.

Mark Cave is Senior Historian with The Historic New Orleans Collection. He created the oral history program at The Historic New Orleans Collection and has developed oral history responses following Hurricane Katrina in 2005, the Deepwater Horizon Oil Spill in 2010, and the Covid-19 pandemic in 2020. He is past president of the International Oral History Association and is co-editor of *Listening on the Edge: Oral History in the Aftermath of Crisis* with this edition's collaborator Stephen Sloan.

Timothy Cooper is a senior lecturer in history at the University of Exeter's Penryn Campus. His research largely explores themes of nature and environment in the social and cultural history of modern Britain. His research makes wide use of life narratives and oral histories as primary sources. He has published on the intersections of environmental and natural disaster in the late twentieth century and is presently completing a book exploring the oral history of the *Torrey Canyon* disaster examining the experience of environmental change "from below."

Deputy Chief Judge Caren Fox is located in Gisborne and is the resident judge for the Tairāwhiti District of the Māori Land Court, hearing cases in Ruatōria and Gisborne, New Zealand. She has acted as legal counsel for treaty claimants and Māori land clients. Fox has also been the presiding officer for the Waitangi Tribunal on the Aquaculture claims, the Te Arawa Mandate and Settlement claims, Central North Island Stage 1 claims, the Te Kohanga Reo claims, and the Te Rohe Potae claims. She is currently the presiding officer for the Porirua ki Manawatu claims. Deputy Chief Judge Fox was also appointed as an Alternate Environment Court Judge in 2009.

Anna Green is an associate professor and currently acting director of the Stout Research Centre at Victoria University of Wellington, New Zealand. She is an oral historian who has published in the fields of labor, community, family, and public history, and while working at the University of Exeter she collaborated with Tim Cooper on the *Torrey Canyon* oral history research project. Another collaborative publication is *The Houses of History*, co-authored with Kathleen Troup.

Heidi Hutner, associate professor of English, sustainability, and women's and gender studies at Stony Brook University, is a scholar of nuclear and environmental history and ecofeminism. She is the winner of Sierra Club Long Island's 2015 Environmentalist of the Year Award. At Stony Brook University, she teaches courses on environmental literature, history, and film. She chaired the Sustainability Studies Program for six years and was associate dean in the School of Marine and Atmospheric Science. Hutner publishes widely on nuclear, environmental, and gender issues. Her work *Accidents Can Happen: Women and Nuclear Disasters* is both a book project and a feature-film documentary. She writes for the *New York Times*, *AEON*, *Longreads*, *Ms. Magazine*, Public Radio International, *DAME*, *Spirituality and Health*, *Mom's Clean Air Force*, *Yes!*, *Tikkun*, and more. Hutner produces and appears in the popular web video show *Coffee with Hx2*, in which she interviews world experts, Nobel Peace Prize winners, McArthur Genius Fellows, and other luminaries on sustainability and environmental issues. Hutner was the associate producer of the off-Broadway climate-change musical *Endangered*.

Debbie Lee is Regents Professor of English at Washington State University. She is the author/editor of seven books, including *The Land Speaks: New Voices at the Intersection of Oral and Environmental History*. She co-directed the Selway-Bitterroot Wilderness History Project, an oral history initiative funded by a grant from the National Endowment for the Humanities. Lee's creative nonfiction has appeared in *Narrative*, the *Montreal Review*, *Terrain*, the *Los Angeles Review of Books*, and elsewhere, and her next book *Remote: A Love Story* is forthcoming with Oregon State University Press.

Suroopa Mukherjee has taught English literature at Hindu College, English Department, Delhi University. She was a fellow at the Nehru Memorial Museum and Library (2005–2008). Her book *Surviving Bhopal: Dancing Bodies, Written Texts, Oral Testimonials of Women in the Wake of an Industrial Disaster* has been widely acclaimed. She is a life member of the Oral History Association of India. She has been staff advisor of a student group set up in Delhi University, We for Bhopal, which brought out an in-house magazine *Hinterland* and made a documentary film titled *Closer to Reality*. She writes extensively as an academic/fiction writer/campaigner/activist, as well as a children's writer. Mukherjee's most recent works are nonfiction adventure stories (a set of four books) based on manmade and natural disasters for children.

Tero Mustonen is a fisherman and the leader of the Kesälahti fish base in North Karelia, Finland. He has been also trained as a scientist (human geographer). He works for the Snowchange Cooperative, a pan-Arctic organization devoted to the preservation and advancement of the endemic, indigenous cultures of the North. Mustonen has worked in Alaska, Canada, Greenland, Iceland, Faroe Islands, Sámi areas, and Finland, as well as the Russian North and Siberia, and New Zealand and

Australia, on climate change, oral histories, and biodiversity issues. He is a speaker and widely published author, including articles in *Science, Biological Reviews,* and other international journals. Between 2018 and 2022, he served as a lead author of the AR6 of the Intergovernmental Panel on Climate Change.

Javier Arce Nazario is an associate professor at the Department of Geography at the University of North Carolina at Chapel Hill, and an adjunct researcher at the Institute of Interdisciplinary Research at the University of Puerto Rico at Cayey. His research focuses on integrating approaches from the humanities, social sciences, and natural sciences to better understand human–environment dynamics and to identify effective ways to translate research beyond the walls of academia. His interdisciplinary endeavors have been recognized by the Presidential Early Career Award for Science and Engineering and the Lowe Art Museum's *Antillean Visions* Jury Grand Prize.

K. Jan Oosthoek is an environmental historian and associate member of the Centre for Environmental History at the Australian National University. He is author of *Conquering the Highlands: A History of the Afforestation of the Scottish Uplands* and editor of *The Globalization of Environmental Crisis* and *Managing Northern Europe's Forests: Histories from the Age of Improvement to the Age of Ecology.*

Dai Qing, a Chinese writer, is renowned for her fearless and incisive work on China's environmental and political issues. Her numerous books include *Yangtze! Yangtze!* about the costs of the Three Gorges Dam, and *Deng Xiaoping in 1989*, the untold behind-the-scenes account of the June 4 Tiananmen massacre. Though her works are banned in China, Dai Qing continues to live in Beijing and publish around the world. She is the recipient of many awards, including the Golden PEN for Freedom, the Goldman Environmental Award, and the Harvard University Nieman Fellowship. Originally trained as a ballistics engineer, Dai Qing is one of China's most important historical investigative journalists.

Christopher Sellers is a professor of history at Stony Brook University. His research concentrates on the history of environment and health, of cities and industries, and of inequality and democracy, with a focus on the United States and Mexico. He is the author of *Hazards of the Job: From Industrial Disease to Environmental Health Science*, *Crabgrass Crucible: Suburban Nature and Environmentalism in Twentieth-Century America*, and is currently completing studies of race, inequality, and environmental politics in Atlanta, and the oil industry and the environment in Mexico and the United States. Since the late 1990s, his scholarly projects have drawn deeply on oral histories.

Jan Bender Shetler is a professor of history at Goshen College. She conducted most of her field research in oral history in the Mara Region, Tanzania. Her work has explored the history of social memory, identity, environment, and place. She has edited a number of collections of locally written histories, including *Telling Our Own*

Stories: Local Histories from South Mara, Tanzania. Her work in environmental history includes *Imagining Serengeti: A History of Landscape Memory in Tanzania from Earliest Times to the Present*. She is working on developing the Mara Cultural Heritage Digital Library to make these interviews accessible particularly to people in the region. Her *Claiming Civic Virtue: Gendered Network Memory in the Mara Region, Tanzania* tells the fuller story of women's oral histories.

Stephen M. Sloan is the director of the Institute for Oral History and an associate professor of history at Baylor University. He teaches environmental history and is an author of several articles, chapters, and books on aspects of oral history methodology, including the co-edited, award-winning *Listening on the Edge: Oral History in the Aftermath of Crisis* with this anthology's collaborator Mark Cave.

Kang Xue is a retired journalist who worked for 33 years with the Beijing People's Broadcasting Station. She has long been concerned about and reported on China's environmental problems, especially water issues. In 2001, Kang Xue edited the book *In the Name of Life - Reflections on China's Water Resources Issues*. Beginning in 2010, Xue organized and wrote a series of "Oral Chinese Water Stories" for English readers. In the 1990s, her enthusiasm for public welfare led her to become involved in the events of the Friends of Nature, China's first nongovernmental environmental protection organization. Xue is currently a member of the Friends of Nature's board of directors and participates as a volunteer in the "Nature Education" and "Low-Carbon Family" projects of the Friends of Nature.

Introduction

Querying Environmental and Human Landscapes

Stephen M. Sloan

We truly live in an environmental moment. From extreme weather events and the effects and legacies of industrialization to the increasing demand for energy and the challenges of global pandemics, there are few issues—political, economic, social, or cultural—that do not connect to the human relationship with the environment. It is a period in our collective history where there is a deep desire to better understand the interplay between individuals, communities, and landscapes—each facets of complex modern ecosystems. In this era defined by globalization, environments rarely operate in isolation but are routinely caught up in an intricate network with places and issues both proximate and distant. For environmental oral historians, this deep need to understand the dissonance, harmony, and all the variances in between in our connection with environments has led to the creation of innovative and groundbreaking research projects. Oral historians now work around the globe, in diverse settings, to explore the wide range of human exchange with the environment.

The emergence of both oral and environmental history as approaches in the second half of the twentieth century were fueled by kindred motivations. Central in each was a growing realization that the dominant frameworks for understanding past events or historical developments were insufficient. In studying the environment and in using oral history, scholars sought methodologies or pathways to examine often overlooked or underrepresented perspectives. For pioneering historian Roderick Nash, studying the environment offered history "from the bottom up," drawing from "the biota and the land itself."[1] These scholars saw landscapes as more than a passive backdrop to the human past and posited that these features should have a central place to truly understand both the past and the present context.

Stephen M. Sloan, *Introduction* In: *Oral History and the Environment.* Edited by Stephen M. Sloan and Mark Cave, Oxford University Press. © Oxford University Press 2022. DOI: 10.1093/oso/9780190684969.003.0001

As environmental historians went from the soil up in researching and understanding historical phenomena, oral historians worked to elevate the most fundamental societal element—the individual perspective—to comprehend a lived experience of the past. These practitioners argued that it was through gathering a close record of the human experience, captured firsthand through oral history, that they could achieve fresh insights and expand the dimensions of their research. Environmental and oral history each embraced this impulse to learn from the ground up, highlighting overlooked settings and discovering new narratives. Their work shifted understandings and brought new meaning to a host of broader events, issues, and phenomena.[2]

For both oral historians and those who studied the environment, however, "bottom-up" approaches were about more than just amassing additional analytic perspectives. It was about addressing deep biases or silences in the historical record. Practitioners saw something essential was missing. As environmental historian Donald Worster described it, the rise of environmental history was "part of a revisionist effort to make the discipline far more inclusive in its narrative than it has traditionally been."[3] This motivation was central in much of the work of oral historians as well. Although there were early debates on the merits of interviewing "elites" versus "non-elites," the popularization of oral history came with gathering "recollections of soldiers and antiwar protestors, women in the war industries, Holocaust survivors, civil rights demonstrators, political refugees, immigrants, and those grappling with their sexual identity."[4] The great capacity of oral history to allow people to speak for themselves, rather than be folded into more traditional top-down narratives where others often spoke on their behalf, was well demonstrated in valuable oral history work on these subjects.

Related to these early motives, the environmental and oral history movements were also deeply driven by an activist desire to push for social reform and respond to societal crises. Early oral historians emphasized working with oppressed communities to gather individual experiences, to try to ascertain community needs and, from that, map and discover the dynamics of structural oppression.[5] Spirited work by oral historians has undergirded progressive social action and movements for social justice for generations now. A flowering of this effort came with the popularization of oral history in the 1960s and 1970s. Similarly, high-profile crises and concerns spurred the growth and maturation of environmental history as a novel research approach in the same era. A landmark of this development in the United States was the collective faculty response to the 1969 Santa Barbara oil spill that created a new interdisciplinary environmental studies program at the University of California, Santa Barbara.[6]

This compulsion to respond to disaster in North America would be reflected internationally, as environmental history research grew from responses to ecological crises in Europe, Latin America, and Asia in the second half of the twentieth century.[7] On occasions where public concern for the environment spiked, such as European alarm following the 1986 Chernobyl disaster, it was connected to increased calls to understand broader issues and the historical context for such events.[8] As Bao Maohong declares, "Serious environmental problems need a historian to offer an explanation," and, as in the case of his home country China, traditional history education did not concern itself with the relationships between people and the environment, so the need is even greater in such settings.[9] Likewise, oral historians have long sought to respond to crises as well, whether they be societal or environmental, and have followed a kindred impulse to document and understand phenomena in the immediate aftermath of events.[10]

In their formation and growth, environmental and oral history also both benefitted from roots and relationships that were broadly interdisciplinary; both their practice and theory were watered by tributaries from many frameworks. For oral historians, engagement with a myriad of fields as diverse as journalism, sociology, anthropology, palliative care, documentary studies, and public history fueled a cross-pollination that allowed for regular reconsideration and evolution of the methodology. Instead of sliding into comfortable ruts with well-delineated boundaries, as environmental historian Douglas Weiner notes, "those fields (like our own) with visibly imperfectly delineated, or worse, highly permeable, migratory, or even barely discernable boundaries provide unexcelled opportunities for intellectual conversations and explorations of a scope, variety, and daring rarely encountered in more fixed fields."[11] The original and evolving set of concerns of both oral and environmental history made each approach a big tent that drew from and spoke to multiple areas of study.[12]

Connected to this desire for fresh insights, a varied disciplinary base, and a penchant for innovation, both oral and environmental history have embraced novel tools in their work. For the oral historian, moving from wire recorders to reel-to-reel to cassettes and, finally, expanding into digital recording, marked the growth of the methodology. Many oral historians also opted to use video in their work, and some shifted to telephone and internet-based methods for distance recording. For environmental study, the written record was often not sufficient, and where traditional source materials failed, advanced techniques such as diatom analysis, tree-ring dating, and ice-core studies delivered the needed information.[13] Both oral and environmental history as practice have evolved rapidly with the possibilities provided through the shift from the analog to the digital age. In their work, scholars of environmental and oral history often pushed back against the textual fetish that generally characterized Western scholarship. Central to each approach was a foundational commitment to fieldwork, to going out to both discover and assess data in new ways. As historian Jan Oosthoek wrote, "Environmental history cannot be conducted in the seclusion of the archive reading room or with a one on one indoor interview. One has to get out into the woods and other environments and experience it: see the real thing."[14] Environmental and oral history practitioners embraced the precept that knowledge is "out there" to be discovered and that it is the task of the researcher to take study to the field to find it.[15]

Endeavoring to understand one place or one story, even one that seems unassuming on a surface level, can be a considerable effort. A classic representation of this in environmental studies is John Hanson Mitchell's *Ceremonial Time: Fifteen Thousand Years on One Square Mile*.[16] Here, Mitchell constructs a comprehensive history of Scratch Flat, the folk name of an area about thirty-five miles west of Boston. This examination grasps long-form changes of a small plot of land and the subtle shifts in the human perception over the *longue durée*. For many environmental and oral historians, digging deep in one place is key—peeling back connections and exchanges that may function at several different levels and evolve over time. It is in understanding the history of the particular, an individual or a community's past or a place's history, that more universal themes can be understood. A lens, precisely focused to see a specific case clearly, can then be helpful in examining broader, more universal phenomena. In their interviews exploring the twentieth-century transformation of Kielder Forest in Northumberland, UK, historians David Moon and Leona Jayne Skelton discovered that:

> The changes in the environment in the relatively small area at Kielder over the last century, and the attitudes to them revealed by the oral history project, can serve as microcosm for much broader, indeed global, processes of environmental change. The changes at Kielder, as elsewhere in the world, have been driven by economic and political considerations,

and utilized various, and changing, technologies. In changing their environments, moreover, human societies around the globe have changed themselves.[17]

Through inventive inquiry employing new tools in fieldwork, scholars working in environmental and oral history introduced vital counter-narratives. Lively reappraisals brought forth through work on the environment and through the efforts of oral historians invigorated thinking and understanding on a host of topics over the last few decades. Oral history contributed to shifts in foundational understanding in historical subfields such as African American history, cultural history, community history, indigenous history, labor history, social history, and women's history, to name a few. Environmental historians as well were innovative and dedicated, undergirded with a strong philosophy of why their new approach was needed.[18] Because of this, oral and environmental history became more than just a subfield such as cultural, social, political, or economic history, but instead grew as a method to better understand history in each of those arenas, as well as others.[19]

Despite similar foundations, aims, and features, there were limited cases in which scholars used oral history to study environmental issues in each approach's first few decades. It was not until the turn of the twentieth-first century when studies integrating oral history as part of a research design for studies of the environment emerged. Early environmental oral history research often clustered around incidents of environmental disaster. Whatever the calamity, both environmental and oral historians worked to trace the ways in which individuals and communities faced such events, as well as present the evidentiary value of listening to such experiences. Several of these projects and publications dealt with natural disasters such as hurricanes, volcanoes, earthquakes, floods, and tsunamis. Though not as numerous, impressive efforts also emerged to understand more man-made environmental disasters, including pollution, chemical and oil spills, deforestation, over-farming, and nuclear meltdowns. Each of these studies sought to find the human voice, what it meant to live a life intersected by catastrophe, and what it can tell us about these phenomena. A landmark in this literature was Svetlana Alexievich's *Voices from Chernobyl: The Oral History of a Nuclear Disaster* (1997, English translation 2005). This extensive study traces stories of death and survival, as well as the intricate legacies that formed in the aftermath of the catastrophe.[20] It reflects well the ways that environmental disaster provides a salient point at which the architecture and priorities of a society can be revealed through the powerful lens of personal narratives.

Although impressive work continues to be done on disaster, the second wave of environmental oral history blended in innovative oral history research efforts that explore environmental change through examining ongoing, engaged relationships between individuals and landscapes. This work is well centered in the "human turn" that occurred in environmental history in the 1990s. This was a pronounced shift in the literature to bring people back into historical narratives. This change worked against earlier tendencies in environmental history where, as J. R. McNeill describes, "human agency disappears into the shadows, while climate, or viruses, or technology hog the spotlight."[21] This changing historiography addressed both old and new topics, viewing human activity as part of the story on phenomena such as climate change, landscape restoration, land use change, indigenous landscapes, natural resource management, and conservation.

A representative work in the United States was Peter Frederici's *What Has Passed and What Remains: Oral Histories of Northern Arizona's Changing Landscapes.* This title frames oral histories of longtime residents' experiences of the rapidly suburbanizing landscape of Northern Arizona as a method to understand an era where traditional land use was being challenged considerably. Treatments such as this elevated the role of local knowledge and instilled in it a value that had often been ignored, or diminished, in prior literature.[22] A fine demonstration of the varied forms of knowledge that can come through narratives is seen in Nancy Jacobs' *Birders of Africa: History of a Network,* which draws out three distinct types—vernacular, ornithological, and recreational. Her work layers these over a recent history of Africa to explore a host of complex topics, including "interactions between vernacular knowledge and science, relations among colonial subjects and imperial authorities, the operation of race and honor among people, and the decolonization of colonial cultural and scientific practice."[23] This phase of environmental history embraces what oral historians always recognized as worthwhile: the great worth of local knowledge. New pathways of environmental knowledge have been opened up through fresh intersections of oral tradition, knowledge about the land and landscape, and oral history.

By the turn of the millennium, the broadening of environmental history led J. R. McNeill to categorize three vibrant varieties of study in the field. He argued that environmental histories were material, cultural/intellectual, or political. The first concentrated on shifts in biological and physical environments, the second on depictions and thinking about nature over time, and finally, political environmental history dealing with law and state policy related to the natural world. McNeill noted that "Environmental historians tend to be more comfortable in one or another of these rings, although some are willing and able to bounce around among all three, even between the covers of a single book."[24]

Environmental history had also grown from narratives of domination and decline to studies that saw mixed purposes and meanings being played out in environments. This became a key characteristic of the next phase of environmental history, which embraced hybrid landscapes. For environmental historians, focusing on hybrid landscapes meant giving up the notion of an ahistorical Eden that human activity had sullied. It also opened up, as Paul Sutter notes, the capacity to "find cultural traces in what they thought was the natural but also to find the natural thriving in places swamped with human activity."[25] This shift also sat well with the oral historians, who had long embraced competing counter-narratives in their methodology and the value of nuance and subjectivity over declarative expressions of certitude.

Despite a growing number of environmental studies incorporating oral history, there have been few reflective examinations of what it means for oral history to be used in environmental study. The many titles included in this volume's bibliography have oral history elements, but few attempts have been made to deliberately bring together findings and reflect on the nature, meaning, and development of environmental oral history. Two important anthologies, both published in 2017, are the exceptions—*The Land Speaks: New Voices at the Intersection of Oral and Environmental History*, edited by Debbie Lee and Kathryn Newfont, and *Telling Environmental Histories: Intersections of Memory, Narrative and Environment*, edited by Katie Holmes and Heather Goodall. *The Land Speaks* focuses its oral history excerpts and corresponding analyses to examine the public lands legacy in the United States. As the book's title indicates, the editors argue that "like any human being, the world of nature has agency of its own. It lives, it acts, it responds. . . . It also speaks to those who learn to listen."[26] The book highlights intersections of oral history and environmental history through the diverse landscapes of North America—urban, desert, forested, cultivated, and wild. Composed by two editors from Australia, *Telling Environmental Histories: Intersections of Memory, Narrative and Environment* assembles scholars from Australia, Canada, the UK, and the United States. These contributors draw out the ways in which oral history has been used to study river ecosystems, processes of industrialization and deindustrialization, and gendered perspectives of environmental change in farming communities. A distinctive feature of this volume, especially in comparison to *The Land Speaks*, is its treatment of urban and industrial environments.[27] This title brings in some global elements that offer useful comparative treatments of different but related ecosystems.

In the present book, scholars from around the globe and researching from many disciplinary frameworks present and analyze their work in environmental oral history. The core of each chapter is an excerpt from stories from a wide range of narrators, to allow those individuals to speak for themselves and detail their relationships to the environments that their lives have encountered. They are farmers, wildlife rescue volunteers, activists, environmental disaster survivors, elders, water system managers, indigenous voices, tribal trustees, wilderness rangers, reindeer herders, fishers, and foresters. All of these oral histories are weighty and timely, offering rich perspectives on the lived experience of disparate environments but also the legacies and complexities of our collective relationship to environments. From landscapes desert to Arctic, they touch on critical issues such as drought, chemical leaks, oil spills, nuclear disaster, indigenous control of resources, natural resource management, wilderness, and protest.

The strength here is not just the collection of compelling stories presented, but the manner in which they are analyzed by the experts who did the research and fieldwork. Their expertise assists in drawing out layers of meaning and significance from the stories shared by their narrators. This compilation showcases and benefits from the interdisciplinary nature of environmental oral history. Historians are well represented here, but so are colleagues working in journalism, human geography, English, environmental science, law, literature, and sustainability. They have contributed to analytical discourses in their field by taking questions from their discipline to the environmental and human landscape in their search for answers. Included here are extreme events—disasters such as chemical leaks, oil spills, and nuclear meltdowns—but also examinations of the slow processes of environmental transformation that rely on the ecological or local knowledge found in interviews. In considering McNeill's varieties of environmental history—material, cultural/intellectual, or political—this compilation offers treatments that touch on each of those variations. Here, environmental oral historians take what could seem broad and impersonal—forces such as climate change, globalism, and environmentalism—and crystalize their meaning and veracity through personal stories that show their significance and lived experience. They reveal the texture and complexities of loss and adaptation as well as the ways that theoretical frameworks manifest themselves in real-world settings.

The dramatic growth of environmental study over the past few decades has occurred within an age of rapid globalization. Shifting views and understandings of the environment have not only reflected this change but also helped to fuel it. Environmental studies have been vigorous forerunners of this shift by overturning narrow frameworks bounded artificially by national borders instead of common ecosystems. As societies more fully embrace the notion that places do not exist or operate in isolation, new ecologies of existence enter the landscape of analysis. Even in those areas or cases that seem quite remote or extreme, the relationships between people and their places, the stories that they tell about them, illuminate themes that resonate throughout the human experience. These deep looks at local phenomena are not episodic, but consistently have a lot to say about more universal topics.

The qualities and strengths of oral history make it a powerful tool for the next waves of environmental study. An evolving blend of the approaches is full of promise. The historiographical shift away from narratives of domination and decline to studies that examine adaptation and change opens new opportunities for the complexities and counter-narratives that oral history can contribute. Oral history's proven ability to address issues of race, ethnicity, class, gender, and sexuality also make it an effective methodology to expand the ways in which environmental studies should seek to explore these dimensions more fully in their work. For oral historians, whose work is strongly associated with cultural and social history, the environmental offers a somewhat novel area of study that should be more central in their practice. There are a host of benefits that can result for this growing connection between oral history and environmental study.[28]

Engaging environmental oral history is about learning more about people and the places they find meaningful, but it is also about a desire to inform change. As environmental historian Marcus Hall observed, "There is also the crucial importance of not overlooking the moral purpose that gestated the field. . . . I don't think environmental historians can or should be disinterested."[29] The same can be said for oral history. Environmental oral history should play an important function in the global conversation about the environment and our collective yearning to chart a more sustainable future.

Critical questions remain that demand answers which oral history employed in studies of the environment can help provide. While the accounts in this volume are significant and revelatory, they should also be seen as a proof of concept to do more. As humankind strives to better understand our complex relationships with environments, both local and global, oral history can assist in getting us there. It is an ongoing dialogue, taking place among all elements of the ecosystem.

Notes

1. Roderick Nash, "American Environmental History: A New Teaching Frontier," *Pacific Historical Review* 41, no. 3 (August 1972): 363.
2. For a useful survey of the growth and development of environmental history around the globe, see J. R. McNeill, "Observations on the Nature and Culture of Environmental History," in "Environment and History," special issue, *History and Theory* 42, no. 4 (December 2003): 5–43.
3. Donald Worster, "Doing Environmental History," in *The Ends of the Earth: Perspectives on Modern Environmental History*, ed. Donald Worster (New York: Cambridge University Press, 1988), 290.
4. Donald A. Ritchie, "Introduction: The Evolution of Oral History," in *The Oxford Handbook of Oral History*, ed. Donald A. Ritchie (New York: Oxford University Press, 2011), 5.
5. For a useful breakdown of this, especially as related to other founding narratives of professional oral history, see Daniel R. Kerr, "Allan Nevins Is Not My Grandfather: The Roots of Radical Oral History Practice in the United States," *Oral History Review* 43, no. 2 (September 2016): 367–91.
6. Occurring in January and February of 1969, this offshore oil spill was the largest spill in US waters up until that time. For more information on the founding of the UCSB program, see Mark Cioc and Char Miller, "Interview with Roderick Nash," *Environmental History* 12, no. 2 (April 2007): 402.
7. McNeill, "Observations on the Nature and Culture of Environmental History," 15.
8. Verena Winiwarter, ed., "Environmental History in Europe from 1994 to 2004: Enthusiasm and Consolidation," in "10th Anniversary Issue," *Environment and History* 10, no. 4 (November 2004): 520.
9. Bao Maohong, "Environmental History in China," in "10th Anniversary Issue," *Environment and History* 10, no. 4 (November 2004): 476.
10. For a discussion of the background and expressions of post-disaster oral history, see Mark Cave and Stephen M. Sloan, *Listening on the Edge: Oral History in the Aftermath of Crisis* (New York: Oxford University Press, 2014).
11. Douglas R. Weiner, "A Death-Defying Attempt to Articulate a Coherent Definition of Environmental History," *Environmental History* 10, no. 3 (July 2005): 416.

12. The host of fields the work of environmental history draws from includes anthropology, archaeology, biology, ecology, forestry, geography, and philosophy.
13. Stephen Mosley, "Common Ground: Integrating Social and Environmental History," in "The Future of Social History" special issue, *Journal of Social History* 39, no. 3 (Spring 2006): 917.
14. K. Jan Oosthoek, "The Role of Oral History in Environmental History," Environmental History Resources, July 27, 2015, https://www.eh-resources.org/role-of-oral-history-in-environmental-history.
15. Danielle Endres, "Environmental Oral History," *Environmental Communication* 5, no. 4 (December 2011): 486.
16. John Hanson Mitchell, *Ceremonial Time: Fifteen Thousand Years on One Square Mile* (Garden City, NY: Anchor Press/Doubleday, 1984).
17. David Moon and Leona Jayne Skelton, "Environmental Change: A Local Perspective on Global Processes," in *Local Places, Global Processes: Histories of Environmental Change in Britain and Beyond*, ed. Peter Coates, David Moon, and Paul Warde (Havertown, PA: Oxbow Books, 2016), 207.
18. Jane Carruthers, "Africa: Histories, Ecologies and Societies," in "10th Anniversary Issue," *Environment and History* 10, no. 4 (November 2004): 380.
19. Ellen Stroud, "Does Nature Always Matter? Following Dirt through History," in "Environment and History" special issue, *History and Theory* 42, no. 4 (December 2003): 80.
20. Svetlana Aleksievich, *Voices from Chernobyl: The Oral History of a Nuclear Disaster*, trans. Keith Gessen (Normal, IL: Dalkey Archive Press, 2005). Based on interviews with more than five hundred eyewitnesses, Alexievich's study was first published in Russian in 1997. The 2005 English translation won the National Book Critics Circle Award for general nonfiction.
21. McNeill, "Observations on the Nature and Culture of Environmental History," 36.
22. As Brian Williams and Mark Riley point out, the use of this ecological knowledge has many names in the literature—including traditional ecological knowledge (TEK), local ecological knowledge (LEK), and indigenous ecological knowledge—but all prioritize recording understandings of the environment. See Brian Williams and Mark Riley, "The Challenge of Oral History to Environmental History," *Environment and History* 26, no. 2 (May 2020): 214.
23. Nancy J. Jacobs, *Birders of Africa: History of a Network* (New Haven, CT: Yale University Press, 2016), 9.
24. McNeill, "Observations on the Nature and Culture of Environmental History," 6.
25. Paul S. Sutter, "The World with Us: The State of American Environmental History," *The Journal of American History* 100, no. 1 (June 2013): 96.
26. Debbie Lee and Kathryn Newfont, eds., *The Land Speaks: New Voices at the Intersection of Oral and Environmental History* (New York: Oxford University Press, 2017), 26.
27. Katie Holmes and Heather Goodall, eds., *Telling Environmental Histories: Intersections of Memory, Narrative and Environment* (London: Palgrave Macmillan, 2017), 15–16.
28. One benefit of this continued exchange will be a fuller attention to oral history best practices in environmental study. One key aspect of this is the need for environmental

oral historians to provide greater preservation and access to their original interview work. At this stage, it is rare when interview fieldwork makes its way to an archive or is shared with the general public. See "Oral History Association Principles and Best Practices" at www.oralhistory.org.

29. Marcus Hall, "True Environmental History," *Environmental History* 10, no. 1 (January 2005): 43.

1

Grim Humor and Hope

Australian Oral Histories of Drought

Deb Anderson

Oral histories by Deb Anderson with residents of wheat-belt communities affected by drought in the Mallee, southeast Australia, Sea Lake, Ouyen, Tiega, Victoria, 2004–7

The Mallee Climate Oral History Collection is the product of Anderson's four-year research partnership with Museum Victoria. From 2004 to 2007, she conducted a series of annual recordings on the experience of drought with twenty-two people in wheat-belt communities dotted across the semiarid Mallee. The timing of the project during the millennium drought coincided with a momentous shift in Australian public awareness of climate change, prompting reflexive discussion of the meaning of drought. Interviewees wore several "hats" in life—farming to health work, public service to parenting, local business to education, government science to community advocacy for rural social and environmental sustainability.

These three interview excerpts illuminate the historical and cultural dimensions of climate. More than "drought stories," these life narratives present multigenerational understandings of place, community, labor, and family that are historically linked in rural culture and threatened by rural decline. These stories bear the mark of rural endurance: as the drought wore on, just one interviewee left the Mallee; the rest were determined to continue making a living here, at the inland edge of the Australian cropping zone.

Interviews with Lynne Healy and Gwen Cooke, Mallee, Australia, September 2004, February 2006, and February 2007

Compassionate, lively, and politically engaged, the pair have much in common. Cooke, then in her fifties, and Healy, mid-forties, were born in the Mallee.

This chapter expands upon oral history research published in Deb Anderson, *Endurance: Australian Stories of Drought* (Collingwood, Victoria: CSIRO Publishing, 2014).

Deb Anderson, *Grim Humor and Hope* In: *Oral History and the Environment*. Edited by Stephen M. Sloan and Mark Cave, Oxford University Press. © Oxford University Press 2022. DOI: 10.1093/oso/9780190684969.003.0002

A patch of dryland salting is the sign of an ecosystem in flux near Ouyen—an agricultural hub of some 1,500 people—in the heart of the wheat belt of southeast Australia. *Courtesy Museums Victoria*

Both ended up marrying farmers (Healy laughs, "I said I would never marry a farmer!"). Both worked off-farm in rural women's health, fighting to secure a permanent, locally based, emotional counseling service for a region experiencing high rates of stress, anxiety, depression, domestic violence, sexual assault, and suicide. They shared this story in 2004.

HEALY: We haven't been getting our average. We've been getting well below our average. Well below.

COOKE: I mean, up until two weeks ago, the crop was actually going back into the ground. You could actually see it *going into the ground*—dying. That was two weeks on the Saturday.

HEALY: Yeah. And a whole lull across the community. It's just like, everywhere you go, *Ohhhh*. You know, "How you going?" "Oh yeah"—

COOKE: "Bit of rain'd be alright"—

HEALY: "Yeah, bit of rain'd be alright." "Oh, the crops aren't goin' real well." "Oh well, I dunno how long they're gonna last." *Ohhhh*. It just *consumes—[she looks to Cooke]* doesn't it—conversation.

ANDERSON: When you're just waiting for rain to come, or praying, how would you describe it?

COOKE: Tense. Extreme tenseness.

HEALY: And every day it gets tenser, doesn't it?

COOKE: I think it was . . . the Saturday the 27th of August, I think it was, that I drove to Mildura.

HEALY: The Saturday?

COOKE: I think it was the Saturday.

HEALY: Yeah, that was the day of the [football] final over here [in Ouyen] and it was stinkin' hot.

COOKE: It was *stinkin'*, stinkin' hot. And just driving up there and seeing those big cauliflower clouds. And I paint, so I'm very aware of what's going on around me. And these big clouds and I—I truly was—I was saying, "Oh, come on God, *please*? Today please?" *[chuckles]* . . . I mean, that's probably the extent of my religion. But I do every now and again say, "Come on, please?" . . .

HEALY: I was at the football that day—*[she looks to Cooke]* and you could imagine. Farmers were there. And it was so hot and I sat there—and I hate the heat anyway—[thinking] *I don't know what I'm doin' here*. But I had a hat on and I kept putting sunscreen on all day. And sitting there and I thought, It's just terrible. Here they are playing football, and it's a winter sport, and it's stinkin' hot. And everywhere you went, the men, you know . . . if you get a drink at the bar or anything like that, the men are all around. And the whole conversation is just about rain, or lack of rain. . . .

COOKE: I'd been up there [to Mildura] four days before, and the change in the crops as I drove along the road was just incredible. They just, you could actually see that they'd *gone* back into the ground. And I truly was in Mildura that afternoon and somebody said, "How's the farm going?" And I could've burst into tears. I just said, "If we don't get rain today, we're *stuffed*. Absolutely." And do you know *[chuckles]* that it rained that night? We got forty points the next morning! *[chuckles]* . . . I'll never forget it. . . .

HEALY: It's like, it's like a *tease*.

COOKE: . . . I went out and there was ten mils of rain in the gauge . . . and I couldn't throw it out! *[All laugh]* It was just lovely, that ten mils of rain!

HEALY: Well . . . I was lyin' in bed and I wanted to get out, strip off and run around the backyard naked. *[Ripples of laughter]* I did! *[All laugh]* . . .

But that's how . . . I just . . . I went to bed early the other night, because I just wanted to listen to the rain. I turned everything off; turned me wireless—I always listen to the wireless at night—

COOKE: You turned it off—

HEALY: —Turned everything off and just lied there and just listened to the rain. . . .

COOKE: I don't think I'll ever forget that 10 mils of rain. I just, I took it out a coupl'a times! *Oh!* I could've drank it—the spider floatin' in it and everything! *[laughs]* It was just so good. And then it rained again, we ended up with ninety points of rain that weekend, which has *really* saved us.

HEALY: Well the feeling I had when it rained that night, it's just like this *glow*. It's like a glow. Internal glow.

COOKE: It's just unbelievable.

HEALY: And it's just like, *Ohhhh! [Cooke laughs]* It is really amazing. Because you don't see it or experience it very often, it's this inner glow, I explain it as: inner glow. And I'm lying there in bed—and it's not my husband beside me! *[laughs]* It's listening to that rain and sitting there thinking, *Oh, just keep coming*. And then you hear it lighten off a little bit and, you know, *No, no, no! No!*—

COOKE: —*Oh no, no, no! [laughs]*—

HEALY: —*Please, please, please keep going!*

COOKE: It's exactly right! It's: *Oh no, no . . . Oh! It's started again! [laughs]*

HEALY: Yeah, then it gets heavier and you think, *Send her down, Hughie! Keep her coming!*

Interviews with Yvonne and Robert McClelland and Bev Cook, September 2004, February 2005, February 2006, and February 2007

Each of these interviewees had lived all or nearly all of their life in the Mallee and had spent the past quarter-century giving countless hours in community service—including to groups established to mitigate the socioeconomic effects of severe drought, such as the Mallee Crisis Committee, formed in the 1980s. This interview is from 2004.

Y. MCCLELLAND: No water in the lake now. . .

COOK: Green Lake, it's been out of commission now for—*[she looks to Robert]* what, five years? . . .

R. MCCLELLAND: It's never been dry this long in anyone's memory. . . . It dried up apparently in the '40s, but it was refilled again. . . .

"We are in climate change now."—Fourth-generation Mallee farmer Robert McClelland says he accepts the climate science, which predicts that there will be more severe and frequent droughts in the dryland Mallee of southeast Australia. *Courtesy Museums Victoria*

COOK: That's right. That's when they used to put corrugated iron sheets up against the railway line, so that the trains could go backwards and forwards—

R. McCLELLAND: Blow the sand off the line.

COOK: Yeah. Oh, just awful! Houses, if they were out in the paddock and with no trees or things, there'd be sand . . . up to the windowsills. You know, just the constant blow. My husband [Jack] said that they'd have a dust storm and you'd put the light on in the middle of the day because you couldn't see. And you'd sit on the verandah, because they didn't have air-conditioning of course in the '40s, and we had a passage right down the middle of our house and he said you used to sit there and just get a bit of breeze. And when you got up, he said, everywhere—except where you sat—was just sand! *[Robert: chuckles]*

And you'd have your meal and instead of having your tablecloth underneath [your plate], you'd have the tablecloth on top. You'd sorta be tryin' to eat your meals under the tablecloth because of—

R. McCLELLAND: —Keep the sand out of your food?! Oh dear! *[chuckles]* ... That's worse than I've heard before too!

COOK: Oh, just this last twelve months we've had a couple of real good ones—

Y. McCLELLAND: —One in June or something, didn't we?

R. McCLELLAND: Last year—June, yeah—there was a bad one. That convinced a lot of us [farmers] to go no-till [a cropping method that reduces soil erosion] ...

Y. McCLELLAND: But I can remember one bad one we had. ... We used to live a few miles further out towards Swan Hill. The house had four paddocks around it that, every three years, used to all be in fallow.

COOK: All be in *fallow*?! *[chuckles]*

Y. McCLELLAND: And this particular year, there was one night we actually packed the kids all into the car and brought them into Robert's mum and dad's to sleep, because it was just impossible to put them to bed out there. ...

COOK: The dust ...

R. McCLELLAND: The sort of aware pioneers left quite a lot of native timber to the west of their homes. And they'd be relatively okay. ... You would always be protected.

Y. McCLELLAND: Ours wasn't! *[chuckles]* ...

R. McCLELLAND: But Joe Watson—*[he looks to Cook]* who'd been up your area somewhere before he went to the war—he said he went to the war and "you could drive up a road," and he came back after the war and, he said, "you could only see the tops of the trees sticking out of the sand!" *[laughs]*...

But anyone whose family has farmed in the Mallee over a long period of time, everything comes true sooner or later. My son's grown up in a very bad economic downturn time. Whereas I was too young when my father had his bad economic downturn in the '30s, and I only just heard about it. ... And then ... we almost had a golden period for twenty or thirty years of good economic conditions and good years. ...

Y. McCLELLAND: Oh and then there was the wool boom in the '50s. Had a big effect on it too ...

R. McCLELLAND: So we knew that you had bad times *and* you had good times. ... There was lots of farmers walked out of the Mallee in the '40s—because after all the Depression years, and then droughts, and shortage of labor in the wartime and everything, people thought the Mallee was finished. Whereas my Dad said he'd heard it all before from his father, 'cause his father was one of the pioneers—started in 'bout 1890—and had had good years and droughts and downturns and survived it all. So my dad thought, Well the Mallee's not finished, the good times will come again. And they did.

ANDERSON: How often do you expect drought?

R. McCLELLAND: Drought? Probably roughly every ten years. But occasionally—like 2002, and the one we've just had [2003] was close to a drought—you get them close together. [In] '82, we were followed by a very good year in 1983, and then '84 and '85 were very poor years. . . . We had a series of poor years—, *[he looks to Cook]* didn't we—*and* poor prices *and* high interest rates. . . . And so it was probably the worst period up here since the Depression in the '30s. We'd just had such a good run for so many years [in the 1970s] that people became very comfortable. . . .

COOK: Thirteen or something years were really, really good. People bought up [land] and they bought machinery and even bought aeroplanes I think! *[chuckles]* . . . I mean to say, people had this mindset that because of all the modern technology, they were just *home and hosed*, safe from anything. But when it stopped raining, well, you don't grow anything.

R. McCLELLAND: . . . And the price of land went up. I think the banks were very . . . *[he looks to Yvonne]* was it the time they were . . . ?

Y. McCLELLAND: Oh! They were *pushing* money onto you.

COOK: Yeah, that's right.

R. McCLELLAND: They were deregulating the finance systems. And the banks were going out actively trying to lend money—and lend money on overseas loans, because their interest rates were lower. And then when the exchange rates went against people, all of a sudden they weren't cheap loans any longer. A lot of people got into trouble in a big way. . . .

COOK: In the '80s, the interest rates went up to 20–23 percent. . . . Some people were paying up to 25 percent.

R. McCLELLAND: There was a number of people who'd received letters from Rural Finance, just saying they were no longer viable . . . Various different people were getting the same letter, but they never knew that *anyone else* was getting the same sort of letter. And they were just devastated! . . . In the immediate thing, they were just frozen. Apparently the chaps—well, what we'd called in our days "depression," but it wasn't talked about as "depression" back then—they just couldn't make themselves go out and do any work, or anything like that.

COOK: Or if they did, they didn't communicate with anybody.

R. McCLELLAND: And I remember there was a big fear that there'd be suicides and things. And because, I think, of all the help that was done, we didn't actually have any. . . .

COOK: There was a lot of the farm women, which was most unusual, but farm women could go over and apply for unemployment benefits—even though they didn't have to say they would work off-farm. And that was probably one of the major things that saved the Mallee, was that at least they had an income that put food on the table. . . . The ones that didn't get the unemployment benefit, the likes of Yvonne, she went out to work. Well, she's a qualified nurse.

Y. McCLELLAND: I did. Had to do my retraining . . . and then picked up work. It almost came looking for me, when I finished.

R. McCLELLAND: And I couldn't believe these wages she was getting! *[Everyone laughs]* I thought, What am I doing farming?!

Interviews with Brent and Melissa Morrish, August 2004 and February 2007

For this couple, the farm was not just home; it was "life." They were still dreaming of the day they could come home to live on it. To make ends meet, these newlyweds, then in their early thirties, spent most of their lives crop harvesting across thousands of miles of rural Australia, from southern Queensland down into New South Wales, then Victoria to South Australia. This story was recorded in 2007.

ANDERSON: What do you remember about the past season?

B. MORRISH: Probably not one to remember, that'd be the first recollection of it. . . . It's the first time in eight years that I've been in the contracting business that I haven't taken a machine to Queensland. . . . There was nothing there. The crops just didn't yield at all.

ANDERSON: Where in Queensland?

B. MORRISH: Usually just southern . . . through a lot of the dry country. From there we were lucky; this year we did start the machines at Moree. They grew a tonne to the acre. . . . Ningan, we usually harvest about 16,000 acres at Ningan; we stripped 1,200. Down near south of Griffith . . . They planted 2,000 hectares [5,000 acres] of dryland crops; we harvested 200 hectares of that. Irrigation there, we stripped some of the biggest crops of irrigation that they've ever grown. There was one crop that did yield 9.2

tonne to the hectare. Other than that, we didn't do anything in southern Victoria. We usually do a lot, between 8,000 and 10,000 acres in southern Victoria; we stripped something like about 2,000 acres. Of that, very, *very* few and far between.

It's the first time I finished harvesting before Christmas. First time since I left school that I haven't had to go back to work after Boxing Day. First time in thirteen years. It's a little bit disheartening but yeah, that's the way it's sorta been.

M. MORRISH: We actually had a nice Christmas holiday though for a change! *[laughs]* . . . The only other thing that kept us afloat is, September '05, we took on a contract spraying weeds under almond trees. . . . Fairly constantly spraying . . . seven days of the week, twelve months of the year.

B. MORRISH: Got a guy full-time now, working for us. . . . Without that, we probably, well, we wouldn't be here—simple as that. . . . It's been probably more of a burden to us, with the way we've lived. Like, we haven't hardly been home here [to Tiega]. As dad'll probably tell you, we haven't been here to help him much, but we've tried our hardest. . . . We sorta haven't really had much of a break in two years.

ANDERSON: And you're heading off again?

M. MORRISH: We're off again tomorrow. Well, just head back up and get things organized for summer harvesting. . . . Kick off in about, oh—

M. MORRISH AND B. MORRISH (IN UNISON): —*Ten days.*

B. MORRISH: But getting back to the start of last harvest, I was sort of a little bit naïve. Thinking it was a bit of a drought and, oh, *things were gonna be okay*—there was a bit of work and things like that. And I hopped in the ute and I had to take a machine to Moree. . . . And when you tow those things, you only do about 65 to 70 kays an hour. I'd driven for about seven-and-a-half hours and driven through most of New South Wales. There was some areas there that I've never seen in my life, in my career as a contractor, *failed*. There was nothing. It was *dirt*, barren. There was nothing. There was stock in poorer condition up there that I've never seen before. It was pretty disheartening all around. Ah, that's when it sorta sunk into me that, you know, things aren't looking very good. . . .

ANDERSON: Do you believe in climate change?

B. MORRISH: Oh, up until six months ago, I would have said no.

M. MORRISH: It's more that you hear about it . . . Well, I mean it's always been there and that sort of thing, but now it's because . . . everyone's raving about it.

ANDERSON: Where do you hear about it or read about it?

M. MORRISH: Oh, on the radio when we're sitting in the tractor on night shifts. They have subjects on climate change and that sort of thing and people can ring in. And that's the thing: you get to have a . . . listen to their sorts of views. Yeah, I think definitely that we're all contributors to it. But [as] to how far that we go, there's gotta be better technology . . . But then, yeah, it's a hard one to get around.

B. MORRISH: Through my travels, I've sorta in the last six months spoken to as many older men—the older the better—to try and find out a little bit of history of past droughts and tough times. A lot of them sort of say that it's similar to this year and similar to that year—previous years. *They don't know*. They're not sure. And that's sorta starting to make my mind a little bit doubtful, when the elderly are not certain on it—when, you know, the last probably four or five years, they sort of give you confidence and know that you know things are changing and it's just, "It'll come good. It'll come good." . . .

M. MORRISH: Like: "The sevens are a good year." They usually have a bit of an old tradition, but now they're a bit hesitant.

B. MORRISH: Well, I know a few of my mates of our era, that are still here trying to farm and make a living, and there's doubt starting to come into it—whether it's really worth doing what we're doing. Because really, if the drought of '06 happens again in '07, I would pretty confidently say that the age group of between twenty-five and thirty-five probably won't be here in two years.

Commentary

Cultural engagement with climate is under constant renegotiation, as oral history is apt to reveal. Indeed if, as Australian historian Tim Sherratt writes, "climate and culture create each other across a shifting, permeable frontier," then those words could form a near maxim for the dynamics of environmental oral history.[1] I grew to understand the strengths of oral history in wheat-belt communities located in the driest part of the state of Victoria, the semiarid Mallee.[2] This is a place whose post-colonial history is anchored in a wide-brown-horizon version of Australia's past—a boom-and-bust tale of backbreaking labor, told with grim humor. It is a place bold in magnitude yet subtle in color, where converging plow-lines force the eye to traverse seemingly endless sandy and loamy layers of beige-tan, rust-pink, and grey-brown. During the first decade of the millennium, the searing heat baked the wheat-belt pale. This collection of stories traversed a period that later became known as the "millennium drought," which peaked in 2006–7 and which some scientists deemed the "worst drought in 1,000 years."[3] Concomitantly, a momentous shift in Australian public awareness of climate change and variability occurred. That shift was far from predictable; in the Australian public domain, climate change carries deep ideological conflicts. That shift formed a moment of *big* history—illuminating how climate shapes culture, and offering an inimitable glimpse of human reflexivity over the ways culture shapes climate.

Interviews for this project began in 2004—on the back of the 2002–3 El Niño-related drought, one of the worst short-term droughts in the nation's recorded meteorological history.[4] From the start, doing oral history formed a means to amplify the dialogue between past and present, as well as the significance of the cultural dimension in understanding environmental issues. Perhaps oral historian Alessandro Portelli concretized this idea best in the assertion that oral stories "communicate what history means to human beings."[5] With that in mind, I asked: How do people live with, struggle with, and understand drought as a component of regional identity? In a way I needed to, for when I began my research only a handful of academic studies had been conducted on the lived experience of climate in Australia,[6] in stark contrast to the swell of interest since.[7]

Both scholarly and public debate in the early 2000s came to center upon drought and what it meant for Australia, for identity, on the driest inhabited continent on Earth. The foremost cross-disciplinary anthology on the topic, *Beyond Drought*, argued that Australians urgently need a more sophisticated climate literacy.[8] In spite of growing understanding of the continent in biophysical terms—of its climate as highly variable, of drought as pervasive, and of the devastating consequences of inappropriate land use—drought still signifies "natural disaster." Certainly, discussion of disaster has long shaped Australian drought policy, which some experts lament as "basically crisis management."[9] Its persistence despite government pledges since 1989 to implement a policy model of self-governance is evident in the escalation of drought relief funding.[10] After the millennium drought peaked, in September 2007 the Australian government allocated more than A$1.1 billion to drought relief—with A$150 million in exit grants to help farmers get out of regions deemed to be in crisis.[11]

It cannot be assumed, however, that either relief funding *or* exit grants pave the way for a more sustainable social or environmental future. Bureau of Meteorology records show that the longest span in which Australia has been relatively free of drought is but eleven years.[12] Paradoxically, though, Australian agriculture has continued down a dominant path of productivism. Farmers are locked into an economic and technological treadmill of production and an associated cost–price squeeze, as declines in real prices for commodities on the global market drive farmers to continually increase production and "efficiency."[13] A wider view of the situation is needed here. Amid the pall of uncertainty hanging over rural futures, the impacts of drought should be viewed as part of greater crisis narrative—which geographers Ian Gray and Geoffrey Lawrence have termed Australia's "global misfortune," fundamentally shaped by a colonial legacy, the introduction of exploitative farming practices to a fragile ecosystem, and the increased vulnerability of family farmers in a wider economy.[14]

On the one hand, researchers point out that we must recognize the contemporary realities of social, economic, and environmental decay in rural Australia, countering the idyll of community and rurality.[15] As social scientists Lisa Bourke and Steward Lockie note, since the depths of decline in the 1980s, once-bold statements on rural crisis have become "so commonplace as to seem clichéd."[16] On the other hand, researchers insist that Australians must truly grasp the role of water in the environment—and respond to it in an "Australian way." John Williams, former head of the Land and Water division of Australia's national science agency, the Commonwealth Scientific and Industrial Research Organization, notes that today, some eight generations on from British colonization, "The critical need is not to drought-proof the inland, for that is impossible. It is to myth-proof Australians."[17] *Beyond Drought* editors Linda Courtenay Botterill and Melanie Fisher may well concur. "Ideally," they write, "the term 'drought' itself should be struck from the national language and replaced with 'climate variability'—or perhaps not be the subject of discussion at all!"[18]

In contradistinction, historians have shown that "recognizing the power of myth" can open the door to a more nuanced historical approach.[19]

Thus, my task as a researcher documenting lived experience of social-environmental catastrophe is to understand ways of thinking about the environment within a wider context of social concerns. This approach is concerned first with how meaning is created, rather than simply what the meaning is. In this respect, I view oral history as invaluable for the insights it proffers into language and culture, bringing into sharp relief how stories of the environment function through and as expressions of the self and the real.

Meaningful themes emerged early in my Mallee project—of drought anchored in the celebratory remembrance of past survival, of uncertainty as ongoing changes in the Australian countryside posed a threat to rural livelihoods, and of putative adaptation as local communities sought solace in their own capacity to "endure" tough times. Indeed, rural histories of the Mallee present spirited sagas of "battling" a harsh climate. The region's name derives from bushland considered by the local Indigenous people to be a supporter of life—for the water in mallee tree roots.[20] Yet, European occupation of the Mallee region from the mid-1800s has been largely narrated as a struggle on a "frontier" beset by environmental extremes.[21] Over the past century, drought and its associated impacts (such as dust storms and erosion) have recurrently tempered the triumphalism of Mallee agricultural settlement, at times culminating in disaster ("Is the Mallee Blowing Away?" cried *The Australasian* newspaper in January 1941[22]). "Disastrous drought" has led to bleak periods of exodus—as environmental historian Tom Griffiths notes, of "failure" so distressing that families abandoned their holdings "in the middle of the night."[23]

Yet for those who stayed, and for their ardent agricultural successors determined to stay put, stories of surviving "dry times" have been elevated into historical terms, cemented in regional discourse as an historical material reality. "Drought is a fact of life in the Mallee," said one farmer who shared his story for the Mallee Climate Oral History Collection. "And if you can't farm in a drought in the Mallee, well you have to get out."[24]

To reach the Mallee from the industrial hum of Melbourne, which clings to the southeast coast, means getting *out there*, into the Outback, far from the city lights. To take the 300-mile (or 480-kilometer) trip inland is to experience a descent in both rainfall gradient and human population density. Concurrently, it is to enter a landscape shaped by primary production: a place engaged in ever-larger-scale industrial agriculture, driven by forces of capital accumulation, economic productivity, and growth.[25] Here, about 2,000 Mallee dryland farming businesses produce half of Victoria's annual cereals crop, mostly wheat for export.[26] Up to one million hectares of the region, which covers 39,300 square kilometers or about one-sixth of the state, is put in dryland (or rain-fed) crops each year—fed on an "average" annual rainfall of 12 inches (300 millimeters).[27] In a dry year, Mallee farmers might sow crops on the mere hint of rain.

If it is through narrativity that we constitute social identity,[28] then stories of drought and memories of rain have been integral to the making of Mallee identities. With that in mind, acquiring an understanding of the interviewees' interpretive frameworks is an essential part of environmental oral history. The challenges of doing oral history on environmental issues are many, but in my case, the big challenge was finding a way to unpack a historical battler narrative of Australian rural endurance without erasing local specificity, nor watering down lived experience as talk. The context in which these people "make do" and "go without," when environmental, social, and economic resources are stretched, is very real. Theirs is a struggle for personal, communal, and cultural self-preservation and self-determination, in a field of competing interests over rural futures. In the Mallee, stories of survival that are both symbolic and literal intersect.

Lynne Healy and Gwen Cooke are not only well accustomed to talking of endurance and uncertainty in the Mallee; these women are trained listeners, too. In their work with rural women in particular—and, by extension, with their families—they'd heard stories of stress (of coping with the anxiety and sleepless worry of what happens when Mallee farm communities don't get rain), isolation (in facing the loss of crops, stock, property, family history, and people, and dealing with the shame of it), depression (and its consequences, of sadness, social withdrawal, and suicide), violence (especially within families), and fear for rural futures. "I don't know how many times I've been pulled up in the supermarket," Healy says. "And these people want to talk *then and there* in the corner of the frozen food section." Their laughter at the memory of rain belies that painful reality. Whenever I listen to or read back over their oral history, I think, if they didn't laugh, they'd cry.

Unsurprisingly, an overarching theme in their discussion was not the non-human environment so much as it was lifting the profile of rural health and social work to achieve better mental health service provision and promote community well-being. In this respect, Healy and Cooke's critical concern with the retention of rural community services spoke strongly to my own lived experience of rural Australia, forcing me to question my motivations in doing this project. I was raised on a dairy farm in a relatively remote pocket of north Queensland. On reflection, I can see this was an experience of the ways rural restructuring inexorably transformed class in rural and regional Australia in the 1970s and 1980s, when governments shut down many regional services, often crippling social infrastructure in these areas. In a countryside bereft of tertiary education choices, for example, I left the north in late 1990, just shy of seventeen, to attend university in a big city down south. On reflection, returning to rural Australia in my work carries a heavy burden of obligation. How might contemporary Australian rural people's understandings of culture, climate, and themselves gain admission into mainstream Australian narratives of rurality? I feel as if I am returning to, even relying upon, the lived experience of an Australian culture that, in an increasingly urbanized nation, many Australians do not have.

Any correlation between the interviewees' stories and mine prompts serious reflection on my own interpretive framework. Healy and Cooke seek to expose the continuing incompatibility that exists between intermittent patterns of welfare support, offered in periods of acute social stress, and the genuine needs of community sustainability. I can see that my work is, in effect, utilizing their voices to expose the local effects of Australia's "global misfortune," which includes the subsequent loss of social services in rural areas, and ongoing local demands for social and environmental security. Further, in their daily work, both expertise and experience are considered vital to enabling a community-driven approach. As health professionals, they use the language of social welfare expertise, yet their advocacy is voiced through individual and familial interpretive experiences of the on-farm effects of drought, and the everyday experience of rural decline. Perhaps the parallel is that I am using oral history as a method that values lived experience and "lay" knowledge *as* expertise, reconfiguring the reception of climate knowledge at the interstices of history, biography, culture, and place.

Evidently, lived experience and the memory of it strongly shapes Mallee farmers' interpretations of climate. In the dynamics of imagination and history, the way memories of lived experience are structured—in the Mallee, poignantly, as narratives of endurance—has the power to undercut crisis-ridden conclusions on the future, even amid clear signs of rural decline. This emerged in conversation with Bev Cook and Robert and Yvonne McClelland. In their interview, the experience of droughts in the 1980s formed a historical marker of socioeconomic and environmental crisis in the Mallee. Their story represents a negotiation of tradition, continuity, and change, an ongoing interrogation of what drought has meant before, during, and since that dark period. Revealingly, in each drought context they recall a host of factors—from low broad-acre commodity prices, unfair international trading conditions, and growing farm debt to the withdrawal of government support for rural industries, farmer unrest over high interest rates, and perceived bank aggression to indebted farmers—limiting the capacity of local people to cope with drought.

That said, with gentle and self-deprecating humor they emphasize the significance of history, politics, and economics as determinants of risk and resilience in Mallee farming. In the 1980s, positivist ideas on the presumed autonomy of enterprise expansion and technological advance were exposed in the rubble of drought. In this respect, in their story, it seems they too are "listening on the edge"—like the ranks of scholars in oral history focused on during ecological crisis[29]—making visible the strengths, weaknesses, and realities of Mallee communities that may not be apparent or so easily observable during "good years."

At times like these, the task and reward of recording oral history on the environment seems almost instrumental, by default illuminating the dynamics of storytelling context. Over time, crisis survival becomes the grounds for the persistence of Mallee optimism, embedding a worldview of Mallee lives defined by cycles of advance and retreat—and advance once more. Robert described Mallee communities in a "good year": "Everyone's confident. Spirits are high."

Confidence had ebbed when I returned to the Mallee in 2006. By autumn 2005, about half of Australia's arable land was drought-affected. By mid-May, farmers' hopes for crop yields were fading by the day and economic forecasters were predicting dire consequences.[30] By June, the National Climate Centre released its forecast for little rain in sight.[31] Much as droughts throughout the twentieth century have been apprehended as "unexpectedly severe" in intensity or duration,[32] in 2005 the nation was pronounced as suffering one of the worst droughts on record—"in our history," stated Prime Minister John Howard.[33] Arguably, however, this drought was discursively intensifying as an indicator of a new narrative of social-environmental destiny. By July 2005, the Australian Government's Greenhouse Office released its first high-level strategic risk and vulnerability assessment of climate change impacts across Australia. It found that recent climate events—including widespread drought conditions, heatwaves, bushfires, and storms—were helping to illustrate the nature of key stresses that might be placed on ecosystems under conditions of climate change.[34]

The unveiling of the threat of environmental catastrophe had the potential to paralyze, however. Environmental crisis became news of the day, every day. As historian Theodore Roszak posited, "At a certain point, we may even grow numb and turn off in confusion or resignation." He put this another way, too, which inspired this oral history researcher: "unless the Earth can speak to us of our dependence in ways . . . that engage the heart, what will all the knowledge of all the experts count for?"[35]

Where a heightened environmental and political context compelled me to extend the scope of my fieldwork, follow-up interviews demanded an adjustment in interviewing approach. I used earlier material as a basis for discussion to elicit more stories of the lived experience of drought, and to directly canvass perceptions of climate change. Extending the oral history project revealed the ways people viewed expertise and experience as inherently partial forms of environmental knowledge. It also exposed the core interpretive problems of climate change. The interviewees began reflecting on their earlier observations. Some doubted themselves. Others doubted the climate science. All doubted the politicians, who weighed in heavily on climate change "debate" in Australia. People were reconciling knowledge of climatic cycles, gained through experience, with more abstract ideas received through science and the media about global change. This abstraction of global environmental risk seemed to affect personal identity and people's sense of order and continuity in regard to their own lived experience. Significantly, many of the Mallee interviewees showed little interest in questions about lived experience that could be answered in the abstract.

Vivid, direct experience dominated the Morrishes' drought story in 2006—although they diverged in their apprehension of prolonged drought as an indicator of climate change. Where Melissa seemed to accept the science (reportedly, women are "more likely to be believers"[36]), for Brent, local memoir shaped the boundaries of knowledge that could be accepted as historical truth. He was struggling to reconcile his own practical knowledge and experience with that of older generations, and that process of identification was dynamic. He consistently measured his own experience against multigenerational family stories of "tough times." This represented a search for utility in the very same cultural practices that could threaten family-farm survival. Indeed, hard times constrained Brent's own hopes for farm succession.[37]

Brent relied on the experiential knowledge—what he called "stories"—of older men, of his grandparents' generation. Evident here was a linguistic transferal, too. As I listened to him speak in folkloric terms of weather cycles, the rhythm of his voice, his intonation and narrative delivery seemed familiar. His storytelling style fit the "farm family history" genre, with the vernacular of farming family culture to which I am accustomed. To me, Brent's search for an interpretive framework within which to rationalize uncertainty implied that an understanding of the power of social identity is essential to understanding how attitudes to climate change are acquired and held. By trying to retain a deeper validation of Mallee farming culture, where climatological competence is tied to lived experience, Brent had deferred acceptance on climate change. Even as he spoke like the older farmers, he chuckled that perhaps he wasn't "old enough perhaps to know what a good season is."

Expressions of humor shaped all of the interviews conducted in the Mallee Climate Oral History Project. While retrospective humor derived from perceptions of predicament, grim humor highlighted a conflicted relationship to lived environment and the conditions of livelihood. Intriguingly, a 1960s North American study of wheat farmers found grim humor to be prevalent too, particularly during drought. That study into the perception of drought hazard highlighted grim humor as a "common element" in the stories of farming men in Great Plains—a region that is, in many respects, comparable to Australia's Mallee.[38] Grim humor was so common that American geographer Thomas Saarinen went so far as to deem it as a personality characteristic of the plainsmen. Tabulating instances of humor by area revealed that three of the driest counties had greater numbers of farmers exhibiting the personality trait. To Saarinen, such humor was linked to "the uncertainty of weather conditions in the area," functioning "to relieve the anxiety brought about by facing a fickle and harsh environment."[39]

Perhaps humor can mask the pathos of the scene of a drought-stricken farm landscape. Here the "local" assumes significance as an identity-based response to wider change.[40] In that respect, amid divisive debate over rural futures in Australia, this oral history project offered a chance to reconsider how rural people recognize drought as a feature of the Australian environment—and themselves as instruments of political mediation. At the same time, their voices cannot help but raise questions of the self-preservative power of narrative and oral tradition for rural cultures under threat.

Notes

1. Tim Sherratt, "Human Elements," in *A Change in the Weather: Climate and Culture in Australia*, ed. Tim Sherratt et al. (Canberra: National Museum of Australia Press, 2005), 4.
2. The project was co-funded by the University of Melbourne and Museum Victoria.
3. John Vidal, "Australia Suffers Worst Drought in 1,000 Years," *The Guardian* November 8, 2006, http://www.guardian.co.uk/world/2006/nov/08/australia.drought.
4. Neville Nicholls, "The Changing Nature of Australian Droughts," *Climatic Change* 63, no. 3 (2004), 323–36.
5. Alessandro Portelli, *The Battle of Valle Giulia: Oral History and the Art of Dialogue* (Madison: University of Wisconsin Press, 1997), 42.
6. The notable exception is Daniela Stehlik, Ian Gray, and Geoffrey Lawrence, *Drought in the 1990s: Australian Farm Families' Experiences* (Rockhampton, Queensland: Rural Industries Research and Development Corporation, 1999).
7. Recent works include: Tom Bristow and Thomas Ford, eds., *A Cultural History of Climate Change* (London: Routledge, 2016); and Tim Sherratt, Tom Griffiths, and Libby Robin, eds., *A Change in the Weather: Climate and Culture in Australia* (Canberra: National Museum of Australia Press, 2005).
8. Linda Courtenay Botterill and Melanie Fisher, eds., *Beyond Drought: People, Policy and Perspectives* (Collingwood, Victoria: CSIRO Publishing, 2003).
9. R. Les Heathcote, *Braving the Bull of Heaven: Drought Management Strategies, Past, Present and Future*, Geography Monograph Series (Milton: Royal Geographical Society of Queensland, 2002), 19.
10. Alan Burdon, "Dry Paddocks, Damp Policies: Drought Assistance Strategies and Their Effectiveness" (Canberra: Australian Department of the Parliamentary Library, 1995), 10.
11. Jewel Topsfield and Canberra, "$150,000 to leave the land," *The Age*, September 26, 2007, https://www.theage.com.au/national/150-000-to-leave-the-land-20070926-ge5wpx.html.
12. R. Les Heathcote, "'She'll Be Right, Mate': Coping with Drought – Strategies Old and New in Australia," in *Drought: A Global Assessment*, ed. Donald A. Wilhite (London: Routledge, 2000), 60.
13. See, for example: David Goodman and Michael Redclift, *Refashioning Nature: Food, Ecology and Culture* (London: Routledge, 1991). Also: Sarah Bell, "Constructing Sustainable Rural Landscapes: Oil Mallees and the Western Australian Wheat-Belt," *Geographical Research* 43, no. 2 (2005).
14. Ian Gray and Geoffrey Lawrence, *A Future for Regional Australia: Escaping Global Misfortune* (Cambridge: Cambridge University Press, 2001), 4.
15. Gray and Lawrence, *A Future for Regional Australia*, 3. See also: Stewart Lockie and Lisa Bourke, eds., *Rurality Bites: The Social and Environmental Transformation of Rural Australia* (Annandale, NSW: Pluto Press, 2001); Bill Pritchard and Phil McManus, eds., *Land of Discontent: The Dynamics of Change in Rural and Regional Australia* (Sydney: University of New South Wales Press, 2000).

16. Bourke and Lockie, "Rural Australia: An Introduction," in Lockie and Bourke, *Rurality Bites*, 1.
17. John Williams, "Can We Myth-Proof Australia?" *Australasian Science* 24, no. 1 (2003): 40–42.
18. Botterill and Fisher, "Introduction," 3.
19. Raphael Samuel and Paul Thompson, "Introduction," in *The Myths We Live By*, ed. Raphael Samuel and Paul Thompson (New York: Routledge, 1990), 21.
20. Don Hopgood, "Foreword," in *The Mallee Lands: A Conservation Perspective*, ed. J. C. Noble et al. (Adelaide: CSIRO Publishing, 1989), v–vii.
21. As stated on the back cover of Rhona van Veldhuisen, *Pipe Dreams: A Stroll through the History of Water Supply in the Wimmera-Mallee* (Horsham, Victoria: Wimmera Mallee Water, 2001). See also Doris Torpey, *The Way It Was: A History of the Mallee 1910-1949* (Ouyen, Victoria: Local History Resource Centre, 1986); and Weston Bate, *Having a Go: Bill Boyd's Mallee* (Melbourne: Museum Victoria, 1989).
22. R. V. B., "Drifting Sands: Is the Mallee Blowing Away?" *The Australasian*, January 18, 1941.
23. Tom Griffiths, "Mallee Roots: A Brief History of Victoria's Northwest," *Park Watch* 178 (1994): 22.
24. Greg Brown, interview by Deb Anderson, February 15, 2005, transcript, "Mallee Climate Oral History Collection." Museum Victoria.
25. The Victorian Mallee is a portion of Australian Mallee country, which also covers parts of New South Wales and South Australia and extends across the Nullarbor through Western Australia.
26. Mallee Sustainable Farming, "About Us," Mallee Sustainable Farming, last modified 2021, https://msfp.org.au/about/.
27. Bureau of Meteorology and the CSIRO, "Regional Weather and Climate Guide" (Bureau of Meteorology, 2019), http://www.bom.gov.au/climate/climate-guides/guides/01-Mallee-VIC-Climate-Guide.pdf.
28. Margaret Somers, "The Narrative Constitution of Identity: A Relational and Network Approach," *Theory and Society*, no. 23 (1994): 605–49.
29. See: Stephen Sloan, "The Fabric of Crisis: Approaching the Heart of Oral History," in *Listening on the Edge: Oral History in the Aftermath of Crisis*, ed. Mark Cave and Stephen Sloan (Oxford: Oxford University Press, 2014), 262–74.
30. Misha Schubert, "Cabinet Debates Extra Drought Aid," *The Age*, May 23, 2005, 4.
31. Steve Waldon, Philip Hopkins, and Melissa Marino, "Drought Tipped to Slash Grain Production," *The Age*, June 8, 2005.
32. Brad West and Philip Smith, "Drought, Discourse and Durkheim: A Research Note," *Australian and New Zealand Journal of Sociology* 32, no. 1 (1996): 94.
33. Howard in: Misha Schubert, "$250m Rescue Package for Farms," *The Age*, May 31, 2005.
34. The Allen Consulting Group, "Climate Change Risk and Vulnerability: Promoting an Efficient Adaptation Response in Australia – Final Report to the Australian Greenhouse Office" (The Allen Consulting Group, 2005), https://research.fit.edu/media/site-specific/researchfitedu/coast-climate-adaptation-library/austra

lia-amp-new-zealand/australia-national/DEH-AGO.--2005.--CC-Australia-Risk--Vulnerability.pdf.

35. Theodore Roszak, *The Voice of the Earth* (New York: Simon and Schuster, 1992), 308–9.
36. George Marshall, *Don't Even Think About It: Why Our Brains Are Wired to Ignore Climate Change* (London: Bloomsburg, 2014), 24.
37. At least Brent Morrish was able to contemplate it. Research last decade found Australian farm successors are still likely to be a son, with most daughters excluded from land inheritance. Characteristically, Melissa Morrish joked about that patrilineal custom: "I'm my dad's son that he never had!" See: Elaine Barclay, Roslyn Foskey and Ian Reeve, "Farm Succession and Inheritance – Comparing Australian and International Trends" (Rural Industries Research and Development Corporation, June 2007), https://www.agrifutures.com.au/wp-content/uploads/publications/07-066.pdf.
38. Thomas F. Saarinen, "Perception of the Drought Hazard on the Great Plains: Research Paper" (Chicago, 1966), 112.
39. Saarinen, "Perception of the Drought Hazard on the Great Plains: Research Paper," 111–12.
40. Rosemary McKechnie, "Insiders and Outsiders: Identifying Experts on Home Ground," in *Misunderstanding Science? The Public Reconstruction of Science and Technology*, ed. Alan Irwin and Brian Wynne (Cambridge: Cambridge University Press, 1996), 126.

2

A Pelican in Her Piety

Perspectives on Wildlife Rescue in Louisiana Following the Deepwater Horizon Oil Spill

Mark Cave

Oral histories by Mark Cave with participants in the wildlife rescue operation in Louisiana following the Deepwater Horizon oil spill, Baton Rouge, Lacombe, Youngsville, and Arabi, Louisiana, November 2010 to March 2011

In spring 2010, a British Petroleum oil rig known as the Deepwater Horizon exploded in the Gulf of Mexico, killing eleven workers and causing a sea-floor oil gusher that released an estimated 4.9 million barrels of oil into the ocean. The spill did extensive damage to wildlife and habitats all along the coast of the Gulf of Mexico. In Louisiana, wildlife agents worked to protect animal habitats and capture oiled animals. BP hired private companies such as International Bird Rescue and Tri-State Bird Rescue to establish a bird rehabilitation center at Fort Jackson, a historic military fort approximately seventy miles south of New Orleans. These rescue organizations staffed the center primarily with local veterinarians and volunteers.

Although most aspects of the response to the spill were closed to reporters, the oil company encouraged the media to visit the rehabilitation center. Images of oiled pelicans filled the nearly nonstop coverage of the event on the twenty-four-hour news networks. The pelican, long a symbol of regional identity, became emblematic of the crisis in Louisiana. Images of oiled pelicans were plastered on protest signs and t-shirts motivating local people to take action in the aftermath of the spill. In fall 2010, The Historic New Orleans Collection initiated an oral history project to document not only the scope and organization of this wildlife rescue effort, but also the motivations of those involved. I interviewed agents of the United States Fish and Wildlife Service and the Louisiana Department of

Mark Cave, *A Pelican in Her Piety* In: *Oral History and the Environment.* Edited by Stephen M. Sloan and Mark Cave, Oxford University Press. © Oxford University Press 2022. DOI: 10.1093/oso/9780190684969.003.0003

Wildlife and Fisheries, as well as veterinarians and volunteers involved in the response.

Interview with Bart Siegel, Arabi, Louisiana, December 9, 2010

Bart Siegel was a volunteer at the bird rehabilitation center at Fort Jackson. At the time of the interview he lived alone with his two cats in a neighborhood that had been shattered by Hurricane Katrina. We talked about his love of animals, which was enhanced by sharing his Katrina ordeal with his cats, which led into a discussion of what motivated his response to the oil spill.

CAVE: Do you think animals have souls?

SIEGEL: Sure . . . well, I'll tell you what, whatever we got, they got. You know, I don't know if they have souls or not. That's a pretty deep question. I have different opinions on a different day. But, whatever we have, animals got. I'm sure we're all the same. Anything that lives and breathes and we're all . . . we're all together. There is a common thread between everything that breathes in this world.

Unemployed at the time of the spill, Mr. Siegel watched the seemingly unending coverage of the spill on the cable news networks. His anger over what was happening motivated him to get off the couch and volunteer at the bird rehabilitation center.

SIEGEL: I saw it on the news when it first happened and I didn't quite understand what was going on until they started to show the oil pumping out of the hole every day, every day, day after day, week after week. And every day I watched it, I started getting madder and madder and madder and realized—getting mad and staying home wasn't the thing to do . . . I managed to see a story about people taking care of pelicans that had gotten oiled and that really drove me nuts. So, I jumped in my car and I went down to Venice looking around for this pelican rehabilitation center. And, after a couple of hours, I finally found it. But, they wouldn't let me in because it was a highly restricted area and I had no—any—you know, who am I? I just went down there banging on the door, "Let me in. I want to help." But that wasn't quite good enough . . . they told me to go to a website that had a test, like a test for the safety, you know, they wanted to know my background. It was a background check. So I did it on the internet, and they called me back, and three days later I was there again . . . all of a sudden, I'm an animal rehab agent without five minutes' worth of experience. But I'll tell you, in a place like that, under those conditions, you get experience in a hurry. You've never seen anything like it in your life. Pelican after pelican. Hundreds after hundreds of pelicans coated in oil, suffering and dying. It's something that changes your life.

A helicopter from the Coast Guard came in with fifty pelicans. They came in dog carriers, big plastic boxes with middle doors. Well, each crate had one pelican in it, and we carried them. They were all covered in blankets, so we carried them directly from the helicopter into the facility. And I never even looked at one. And then, after we had finally unloaded all of the pelicans and got them inside the building, I picked up the sheet and looked inside and lost it. I'd never seen anything so horrible.

Well, anyway, the next day—we left them in the boxes overnight—and the next day, we started to take them out of the crates, the dog crates, and bring them to the veterinarian for a quick check to see if they were strong enough to go through the rest of the treatments, you know, these pelicans maybe hadn't eaten in three or four days, who knows. They're dehydrated, they're weak, they're scared, they're horrified, and now we're going to handle them all the time. You know lots of people are going to be holding onto them trying to take care of them. It's—you know, it's got to be very detrimental to a wild creature.

I ended up being captain of a de-greasing station. I applied a chemical called methyl soyate to the birds in a squeegee bottle, and worked the oil through all the feathers, all over the bird's body, from his beak all the way down to his feet, just drenching in oil. They called it marinating, but we were soaking the oil out by thinning it with this chemical. I imagine it's not an oil-based chemical, but it still thins the oil, and it worked a lot better than the original oils, the vegetable oils that they were using.

So after . . . the oil is thinned out, they're put back in their pen, and they stay there for maybe an hour to two to three hours, depending on how many birds are going through, and then they go to the wash bins. The wash bins consist of four different—four different tubs using Dawn detergent and warm water. And a team of three people bathe the bird. One usually holds and secures the bird so it doesn't hurt itself or anybody else working on it, and the other two hose it down and make sure that it, you know, it just—it's okay physically. You pull it—one pulls out a wing, the other uses the water, and one just holds the bird steady and it goes from tub to tub with the—you know, the amount of Dawn each time getting less and less.

Then the birds are put in a drying room and they stay there for a couple of hours until they get mostly dry. They're preening themselves and they're starting to feel a little bit better, and after that we put them in an outdoor pen with a big water tank. And now they start feeling a lot better, and they start swimming and washing and playing and flying around in there, and you know, we could tell which ones are going to live. But every pelican that we washed, in my mind, got through and lived, and got released. . . . And every time I go down to Grand Isle and see pelicans, I kind of feel like, you know, maybe those are mine. So, yeah, there's a connection to nature. You can't help it, especially when you get that close to it. You know, holding a pelican in your arms and feeling its heart beat and looking in its eye, you know, it's like asking you to help it and you do. So you're connected to nature and, man that's a beautiful thing.

Interview with Jeannie Beaston, Youngsville, Louisiana, November 17, 2010

Jeannie Beaston was also a volunteer at the bird rehabilitation center at Fort Jackson. She was a homemaker whose husband worked in the oil industry. Like Bart Siegel, she watched the seemingly endless media coverage of the event on her television, which motivated her to search for volunteer opportunities online. She filled out an application and within a few weeks she was working at the rehabilitation center. She had never seen a wild pelican before.

BEASTON: So once we walked in the warehouse, it was just overwhelming. I mean, they had over six hundred extremely oiled birds. They had one trailer for us to stay, and it slept twelve people. Nine times out of ten we had probably sixteen people, so we were crammed in there. Yes, it was just overwhelming. I just kind of poked my head in the warehouse, knowing that I had to go through the orientation the next morning and I would be starting the next morning, but I wanted to just see for myself. I was just shocked. I was literally shocked at the amount of birds that were there, mainly pelicans at that point, and just extremely oiled, extremely oiled.

The next morning I got up, and just kind of got thrown right into it. The orientation was just basically more of a safety briefing, what to do and not to do, proper clothing, goggles, tie-back. They wanted to make sure that you weren't going to be exposed to the oil, was the main thing, and then they had the wildlife portion of it, too, how to handle the wildlife, don't talk to the media, just little things like that. We weren't allowed to talk to anybody. We weren't allowed to take pictures. Just kind of the dos and the don'ts. That was pretty much it.

At that point we just went downstairs, and they put us in different areas where they felt the need was. So I ended up in triage, which was probably the craziest, hottest, busiest area.

CAVE: Pelicans have such big and pronounced eyes. Did you find yourself looking into their eyes?

BEASTON: They're soul chasers. We called them soul chasers. As a matter of fact, I remember in triage my main job there was actually holding the birds for the vet so she could examine them and do various things to them. I would kind of pull them up close to me and breathe kind of slow, because they were breathing very, very rapidly. They were scared. So they're looking at you and they're looking at the doctor, looking at you to kind of comfort them. So you're kind of telling them, "It's okay. It's okay." Then you take a breath and you can just feel them release, like, "Okay, I think this is going to be all right. They're not going to kill me." You could make a connection with them. They're very, very smart birds, very smart. And you're right, they're soul chasers, is what we used to call them, because they would just look at you with those eyes, like little human eyes.

CAVE: How did it impact you on an emotional level?

BEASTON: You just kind of put a wall up while you're in it. You have to. You just see so much devastation and so much death. I mean, you're euthanizing birds left and right. You're getting birds that are dying before they even get on the table. The first couple of times it happens, it gets to you. Then you just have to say, "I can't let it affect me every single time," so you put a wall up. You have to put a wall up. I guess it would be like a nurse in an emergency room with a child.

CAVE: Do you think animals have souls?

BEASTON: I do. I do. I actually do. I really do. And I believe we're judged. We're judged for how we treat animals on this earth.

Interview with Jack Bohannan, Lacombe, Louisiana, January 20, 2011

Jack Bohannan was an agent with the United States Fish and Wildlife Service charged with protecting the Breton National Wildlife Refuge. He was dealing with large numbers of concerned citizens who wanted to visit the pelican nesting areas to see if they were all right. Wildlife agents were tasked with the often difficult job of capturing oiled wildlife and bringing the animals to the rehabilitation center.

BOHANNAN: I guess what I'm saying is that there's no one else that's going to watch that resource or protect that resource any more than what we could have or what we were doing. And so we have a lot of folks that wanted to have access, but the thing we don't want to have is more of an impact, have more intrusive entrance into those resources, into those refuges when those birds are actively nesting. And so, if I tell someone "the birds are fine" . . . I'm doing an assessment. I'm doing surveys, of these birds—as a refuge manager responsible for that resource, that should be sufficient, that I'm tasked with maintaining it and caring for it the rest of the year. The rest of the time, this just happens to fall within the area that I'm responsible for, and if my staff and I are doing a good job, which we're doing the rest of the year, we're certainly capable of doing this. And we did that, and I think that really early on, it was just folks could not believe there were that many birds gathered in one area and not have an impact.

CAVE: I know this is kind of a dumb question, but how do you catch a bird?

BOHANNAN: Well—it's challenging. I'm a very large fellow, so it's even more challenging. But—we were issued nets, large . . . I don't want to say fishing nets, but that's what they were—which you had to be careful because you don't want to injure the wildlife while you're trying to capture them. So these had a very wide girth to the size of the net. The netting material was not real fine. It was more of a heavier mesh, good stuff. It wouldn't entangle the birds.

And, the thing about it is, you're not going to catch a bird that's oiled unless it's not flying capable. If that bird is fly capable—and that happened to us a few times . . . you'll never catch one of them like that. And the ones that are able to be caught you would walk right up to and just put the net over them. But there were some that took—a little more work . . . it could still move very quickly and still run. . . .

I picked up one bird—it couldn't even hold its wings up, out with the airboat one day, out on the outer edge of the gulf. And that was a seagull—I had to really study it to figure out what kind of bird I was looking at. I don't know where the oil came from. I can't say it came from that event, but it was a heavily, heavily oiled bird. . . . But we got it and got it turned in to the folks that rehabilitated.

BOHANNAN: We utilized our office in Venice, because we're actually on the water. So, we did have boats pull up to our office and we have four docks. We could back a truck, or a van or trailer, which is what they used, right to the dock, and that way, the wildlife didn't have to go through being toted or carried or bounced. . . . [For] The birds—most birds, most wildlife in general, [the move is] more of a traumatic thing for these animals to go through, more so sometimes than the actual oiling event itself. They're not very hardy—they get very worked up just the nervousness of that animal . . . they'll perish. So, it's very important that you treat these birds as docilely as, as carefully as you can. They're very fragile animals.

Interview with Drew Wirwa, Lacombe, Louisiana, January 20, 2011

Drew Wirwa was an agent with the United States Fish and Wildlife Service charged with protecting the Breton National Wildlife Refuge. He grew up hunting and fishing, and it was that interest that had led him to become a wildlife agent. He began with the Wildlife Service in May 2009 and had been a wildlife agent for just a little over a year at the time of the Deepwater Horizon explosion.

WIRWA: Well, the fear was that people would, for an example, chase one moderately oiled bird into a colony of nesting birds trying to—catch that one, and then actually do more harm than good and damage more nesting birds just trying to get the one moderately oiled bird that may end up doing fine on its own. There was a lot of gray area there and a lot of subjectivity. When is it appropriate to get a bird and capture that bird and take it for rehab? When is—when is that point? When is it necessary to do that and when it is necessary to let them be and let them go on their own? Once everybody started to come on the same page that if it's gonna risk other pelicans in other nests just for that one bird, especially in the large colonies, then you just don't do it; you leave it alone. That just took a while, I think, for that to work itself out. Well, as a fish and wildlife service agency, it's our job to protect populations and, yeah, while individual animals are important, biologically speaking, it's the population that you're concerned about. It's those populations and those habitats which support those populations, and so it's not as much prioritizing as it is stepping back and looking at the big picture and what's—what's important. First, the habitat, you have to have the habitat there. Then you want to make sure that that habitat can support those populations. And you know, individual animals, they're gonna—they're gonna be okay. They're gonna be fine. Those individual animals are important but, as an agency, we look at the—at the populations—we're going to make sure we can support those populations.

Interview with Carol Foil, Baton Rouge, Louisiana, December 15, 2010

Carol Foil was a retired professor of veterinary medicine and one of the founders of the Louisiana State Animal Response Team (LSART). She was responsible for organizing the volunteers who served at the bird rehabilitation center at Fort Jackson.

CAVE: Do you think animals have souls?

FOIL: Oh, dear. You know, in veterinary medicine we have this dichotomy of viewpoints that have been given tags. One of them is the animal rights perspective and the other is the animal welfare perspective. From the animal welfare perspective, it says that the animals that we have charge of or that we are affecting by our actions have a right to be—or we should morally treat them well, whereas the animal rights perspective says that a dog is a boy is a cat is a man, that they all have equal value. I definitely don't ascribe to that. So, I'm not a religious person, though, so it's difficult for me to answer the question you gave me.

One of the most interesting things about this that I learned by accident, was that BP was going to be fined for the birds that died. Each species has its own monetary value assigned to it. . . . There's tons of laughing gulls in this world. Laughing gulls, the population's expanding, you know, all over. So I thought that was really interesting. Laughing gulls are worth—I don't know the value, but it might be $10 compared to $2,000 for the most valuable species. I can't even tell you what a bald eagle, if it had gotten oiled, what the fine would have been. Maybe $10,000 a bird. I don't know. But it was quite interesting that there's a monetary value that came out of previous oil spills in other company-induced natural disasters. There's a monetary value for every species of wildlife out there, as far as fines go.

Interview with James LaCour, Baton Rouge, Louisiana, March 9, 2011

James LaCour was the chief veterinarian with the Louisiana Department of Wildlife and Fisheries. He advised animal rescue groups involved in the response on state regulations and represented the state at events associated with pelican releases back into the wild.

LACOUR: As an overall generalization, as organisms, they're very interesting to me. I mean, pelicans, for instance, that was a large percentage of the

A Louisiana Animal Response Team member holding a pelican during a media tour of the Fort Jackson Wildlife Rehabilitation Center. Some of the team members remarked on the emotional impact of looking into the pelican's eyes while holding the animal. *Photograph by Keely Merritt, The Historic New Orleans Collection*

birds, just their anatomy, how they're designed. If you've ever seen one, they have large air sacs in their breasts for when they dive. It cushions them when they hit the water. They can pump up the air sacs in their legs for extra flotation. I mean they're just amazing birds. To sit by when you could do something about that, and just let that bird die, is silly. I think I'm intrigued by the animal itself, if you will.

Commentary

The iconic figure of a mother pelican pecking at her own breast to draw blood to feed her young has been an important symbol of sacrificial love for more than 700 years. Known as *The Pelican in Her Piety*, the image has been widely used in Louisiana because of the traditionally large native populations of brown pelicans in the state. Since the French colonial period, the symbol has adorned Louisiana churches and Masonic lodges.

It has been a common architectural motif on public buildings, and eventually the state adopted it as their official symbol and centerpiece of the Louisiana flag. Over the past century, representations of the pelican have been used to promote almost every type of local business, from a bowling alley to a cracker factory. It was the namesake of the professional baseball team in New Orleans from 1887 to 1959 and is the current name of the city's professional basketball team.[1]

In the aftermath of Hurricane Katrina, which battered the Louisiana coast in 2005 and called into question the future viability of New Orleans and other nearby communities, such long-standing symbols of regional identity took on an even greater significance. The image of the pelican became an important symbol of community pride through the long struggle to rebuild communities. It was understandable then, that in Louisiana, images of oiled pelicans quickly became emblematic of the Deepwater Horizon oil spill crisis. As a symbol of a collective identity, the suffering oiled bird came to represent the destructive effects of neglecting responsibility to the local environment.

Oil exploration and production reshaped the landscape of South Louisiana throughout the twentieth century. More than a million producing wells have been drilled; more than seven thousand offshore platforms constructed; and thousands of miles of pipeline canals have been cut through coastal wetlands, causing saltwater intrusion, erosion on a massive scale, and dramatic impacts on wildlife.[2] Technological advances since World War II have enabled oil companies to search and extract oil from farther out into the Gulf of Mexico. And concerns over dwindling supply since the 1970s have motivated oil companies to take increasingly greater risks to extract oil using deepwater platforms.[3]

The environmental impact of oil exploration, along with damage caused by DDT and other pesticides, culminated in the near extinction of the native brown pelican by the 1960s.[4] The state of Louisiana teamed up with the Florida Game and Fresh Water Fish Commission to relocate 1,276 pelicans from nesting areas in Florida to the Louisiana coast. These new populations were carefully monitored for decades, and their numbers gradually increased to around 24,000 birds. On November 11, 2009, US Senator Mary Landrieu, along with officials from the United States Department of the Interior, made the official announcement with much fanfare at a Louisiana wildlife refuge that the brown pelican had been officially removed from the endangered species list. The successful effort to save the species was a source of pride for many in Louisiana.

Focusing an oral history response to the Deepwater Horizon crisis on the effort to save wildlife, and on the pelican in particular, helped to capture the raw emotional resonance of the spill in ways that longer term environmental impact projects, such as the damage to fisheries and impacts on coastal erosion, could not. It also enabled me to explore how the different individuals involved in the response effort viewed their relationship to the animals that shared their environment, and their sense of responsibility for them.

In the summer of 2010, when the crisis was at its peak, our staff photographer Keely Merritt and I traveled down to the Fort Jackson Wildlife Rehabilitation Center, which, at the time, was located in Venice, Louisiana, near the mouth of the Mississippi River. Later that summer the facility relocated inland to Hammond because of concerns over the impact of potential hurricanes. Those who operated the center gave regular tours of their facilities for the media. Journalists from all over the world were in the region looking for an angle on the story. Most of the action, however, was taking place at sea, or in remote coastal marshes, which were very hard to access. British Petroleum, which was managing the response, was very tight-lipped about what they were doing and blocked media admittance at almost every turn. As a result, journalists flocked to these tours of the animal rehabilitation center. They needed to produce news of some kind, and the bird rescue angle offered them a marketable story with great visuals.

We posed as members of the media and took one of these tours. The tour itself was rather uplifting. Listening to the representatives of the center, we really felt their enthusiasm and dedication to saving wildlife. We were shown plywood pens where groups of oiled pelicans were huddled together awaiting cleaning. Then, we were brought to the washing area where we were given close access to the animals being washed. This point in the tour provided great photo opportunities. Finally, we were taken to an area they called "Pelican Island," which is where the rehabilitated birds were kept. To me, it looked very much like a children's play area. You could not help feeling a little better about the horrible event that was unfolding, and the many news stories produced from these tours reflected this sentiment. For me, it was a great opportunity to find potential interviewees, make contacts, and form a frame of reference which would be invaluable in initiating interviews a few months later.

The organizations that BP hired to operate the rehabilitation center, International Bird Rescue out of California and Tri-State Bird Rescue out of Delaware, are privately owned companies that respond to oil spills. Their corporate partners are almost exclusively large oil companies, and bird rescue enterprises are in competition with one another to secure big oil company contracts. International Bird Rescue is the industry's pioneer. It started operation after a spill in San Francisco Bay in the early 1970s, and it served a high profile role in the response to the *Exxon Valdez* spill in 1989 and in response to the MV *Treasure* oil spill near Cape Town, South Africa, where they rehabilitated nearly 20,000 penguins.[5] Over the years, IBR has developed industry standards for rehabilitating oiled wildlife. The people I talked to from these companies are undoubtedly animal lovers dedicated to helping birds, but the businesses do depend on the oil industry for their continued existence.

Although it is difficult to question the motives of these rescue companies, it was clear that, from BP's perspective, the unspoken purpose of the rehabilitation center was to move the narrative of the crisis away from the damage that was still being done to the environment to one of renewal and rehabilitation. By closing media access to almost all areas of the response except the rehabilitation center, BP could move the narrative in a more positive direction.[6] Images of rehabilitated pelicans and descriptions of the process used to save them were irresistible to the media and were voraciously consumed by the public. What the media was not shown was the condition of the animals when they were first brought in, the euthanasia of animals that were too far gone, or the heavily oiled carcasses of dead birds, which far outnumbered those who could be rehabilitated.

Although International Bird Rescue and Tri-State Bird Rescue managed the center, it was staffed primarily by what the oil company called "volunteers." In reality, they were paid, about $200 a day for the work that they were doing. Paying the volunteers likely helped BP build a more stable and reliable work force, but it also gave the oil company some leverage in controlling the workers' interaction with journalists. Jeannie Beaston, for instance, noted several times in her interview that she was not allowed to speak to the media. Bart Siegel, who was an amateur photographer, was told to not take pictures in restricted areas at the center. Confrontations over this policy eventually led him to leave his "volunteer" post. In addition to giving BP a means of control, calling the employees volunteers helped nurture the feel-good narrative that they were trying to establish through the rehabilitation center. BP employed the rehabilitation center to try to shape the narrative of the spill through the media; voices from those involved in the response which might derail the storyline of this preferred narrative were quickly silenced.

Most of the workers were members of the Louisiana State Animal Response Team (LSART). The mission statement of the organization defines LSART as "an organization of groups and individuals with an interest in animal well-being related to emergencies or disasters."[7] The organization really came into its own following Hurricane Katrina in 2005 and, by the time of the BP spill had a workable online presence, a perfect vehicle for the rehabilitation center to process volunteers. These volunteers not only provided much of the labor used at the rehabilitation center, they also manned remote "stabilization" sites that took in oiled birds from wildlife agents who were capturing them. The membership of LSART was diverse; it included veterinarians, students with an interest in wildlife rehabilitation, enthusiastic bird watchers, and those who simply identified themselves as animal lovers such as Bart Siegel and Jeannie Beaston.

In my interviews, I wanted to get a sense of how each person saw their place in nature, and how this perceived relationship motivated them to take action. I placed a heavy emphasis on life story in the interviews, particularly their interactions with animal life. We talked about family pets, hunting and fishing trips with their fathers, and vacations they took to admire wildlife. I was very attentive to the language interviewees used to describe animals and what this said about either their relationship to nature or their need for emotional distance during a crisis.

A question that I liked to ask interviewees, although I was concerned about the spiritual framing of the question, was if they thought animals had souls. Narrators often greeted the question with a roll of their eyes, but it really forced each individual to think about, and define, their relationship with animal life. In LSART volunteer responses to this question, there was a marked difference between the answers provided by those who defined themselves as animal lovers and those who identified as professional veterinarians. Animal lovers responded yes to the question, or if they were not particularly religious, they responded as Bart Siegel did "whatever we got, animals got." Siegel also mentioned that there is a "common thread between everything that breathes," reflecting a kinship in relation to animals. In the life story portion of his interview, Siegel relates his experience during Hurricane Katrina in great detail. This is common in life stories of people living in and around New Orleans, but he told the story particularly because it informed the discussion of his relationship with animals. At the time of Katrina, Siegel lived alone with his two cats Romeo and Buster. He stayed during the storm, and as the waters started to rise, he grabbed the cats, threw them in a suitcase, and climbed to the roof. He remained on the roof, clutching onto them for the remainder of the storm and for seven horrific days until they were able to leave the city. He remarked, "I took them with me. That's all I took with me. My whole life was under the water. The only thing that was around that was mine was my two cats. I saved them and everything else was gone." This shared experience seemed to have created unique value in his connection to animals.

Jeannie Beaston had a much more maternal view, and one that suggested a strong religious belief. She emphatically stated that yes, animals do have souls, and that we are judged on how we treat them. She lived in the country with her husband and a rescue dog named Boudreau. She talked about when they first moved to that area, when there were hardly any birds in the neighborhood, but she started to feed the wild ducks and now the area was full of birds. When asked why she volunteered, she remarked, "my heart was just breaking, knowing how many animals especially birds, were going to be affected by this." Throughout her interview she compared the pelican to a child, and she remembers holding the animal close to her breast so that it could feel her breath and sense that everything was going to be okay. She also recalled how she and some of the other volunteers cried when they had to euthanize a pelican with a broken neck, probably the result of capture and transport to the center. She remarked, "I just felt so horrible for that poor baby."

Although some of the veterinarians that I interviewed seemed put off by the spiritual framing of the question, they all believed that we had a moral responsibility to take care of animals. Carol Foil grew up exploring the outdoors with her father, who was an amateur naturalist. She became fascinated by birds, and remembered constructing plastic models of birds as a child and hand-painting them to resemble different species. Her dream was to become an ornithologist, but out of practicality she became a veterinarian. When asked if she thought animals had souls, she very analytically broke down the differences between the animal rights perspective and the animal welfare perspective. She pointed out that she definitely did not think that people and animals have equal value, and qualified her response by saying, "I'm not a religious person, though, so it's difficult for me to answer the question you gave me." She questioned why the response effort paid for the transport of laughing gulls from Louisiana to Florida: "There's a lot of laughing gulls in this world. Should they be put on an airplane and flown to the east coast of Florida? I have no problem with the brown pelican because it just came off the endangered species list, but I thought that was quite interesting. I don't criticize it; I just wonder why?"

In her interview, Foil talks at length about the cost of rehabilitation per animal, and whether or not it was worth the cost to save animals that had sufficient populations. She seemed to have a very pragmatic approach to wildlife rescue, and, like the other veterinarians that I interviewed, maintained an emotional distance from the plight of individual animals. James LaCour expresses this emotional distance eloquently in his scientific description of the pelican.

The wildlife agents distanced themselves even further from the suffering of individual animals. Their primary interest was the protection of habitats. Jack Bohannan and Jack Wirwa were responsible for the management of three national coastal wildlife refuges, Bayou Sauvage, Delta, and Breton. Breton, in particular, contained important nesting areas for the brown pelican.[8] The importance of these nesting colonies was well known to bird watchers and nature conservancy groups, and many well-intentioned bird lovers wanted to get out to those areas to see if the pelicans were ok. Both Bohannan and Wirwa discussed their concerns over the impact that these animal activists might have on the habitats, and, at least for Bohannan, some resentment over the implied suggestion that the wildlife agents were not taking adequate care of the animals. In their interviews, agents expressed that the effort to chase down and capture individual oiled birds was not an effective use of their time. They felt that it took them away from their primary purpose of protecting the habitat by laying and relaying boom to keep oil from getting onshore.

The wildlife agents whom I interviewed often grew up hunting and fishing and enjoyed working outdoors. Frequently, that is what led them to be wildlife agents. They worked regularly with hunters and anglers and were often very much a part of that culture. The language that they use in the description of animal life is markedly different from the language used by the volunteers I interviewed. Bohannan, for instance, referred to the pelican population as a "resource" and mentions that he was doing "assessments" and "surveys" of that "resource." This language certainly creates an emotional distance. In part, this language is professional jargon. I conducted the interviews with the wildlife agents at their workplace during the workday, where it is common for people to adopt a jargon that provides distance from the work they are doing. In this case, however, the term "resource" does reflect an entirely different view of wildlife than that held by the volunteers I interviewed. In this view, the individual animal's life has significance only in connection to the population or the "resource." Attempts to save an individual animal's life were not seen as a moral responsibility or even a priority. What was important was the protection of the habitat itself. This perspective certainly does not reflect a kinship with animals, and the habitat that they are protecting is something very much set apart from everyday human experience.

Agents of the United States Fish and Wildlife Department and the Louisiana Department of Wildlife and Fisheries were the primary groups charged with the collection of deceased wildlife. They would bring the dead animals to the rehabilitation center to be stored in freezer trucks. As Carol Foil mentions in her interview, BP was responsible for paying reparations for the animals killed because of the spill. In determining the reparations, each animal species was assigned a particular monetary value. These values, based on reparations paid out for previous spills, were usually based on the animal's rarity. A brown pelican in Louisiana, for instance, was a high-value animal since it had just recently been removed from the endangered species list, whereas a laughing gull or a gannet would be a low-value animal. Although the amount paid out in reparations differed substantially from species to species, the cost of rehabilitation, as Foil remarked, was nearly the same for each of the bird species. This would lead some to suspect that decisions on capture and rehabilitation might have been made on the basis of species. There certainly would have been a significant economic incentive to focus the effort on pelicans and neglect the "low-value" animals. From the perspective of a visitor to the center, it did seem that pelicans were the only birds that were receiving treatment. Those that I interviewed who worked at the rehabilitation center were adamant, however, that species was never a factor in their decisions on which animals to rehabilitate. As Foil stated, "every animal had an equal value, one to the next, to the people in charge at Tri-State and IBR. Most of them are veterinarians that began those organizations. We could say that's the veterinary perspective or wildlife rehabber perspective versus the population biologist perspective. . . . Every individual animal counts just as the next one does."

This notion of high-value and low-value animals, however, is interesting to explore. The pelican was a high-value animal in the hearts and minds of local residents, not necessarily because it was rare but because it was a symbol of collective identity, and, as a result, triggered a greater empathetic response. The plight of laughing gulls, although they suffered as much as the pelicans, did not inspire the same response as suffering pelicans, possibly because the media attention was focused primarily on pelicans, but also perhaps because they did not have this same connection to local collective identity. Just as in a disaster affecting human beings, the life of a person from one's own community or country seems to mean more than a person from another culture. Oral history can provide us a window into how these value judgments are made.

What enables us to feel empathy or compassion for creatures other than ourselves? For Bart Siegel, suffering though the aftermath of Hurricane Katrina with his two cats helped nurture a bond with animals. This belief that "there is a common thread between everything that breathes in this world," as he put it, led him to reach out to help suffering animals after the spill. Jeannie Beaston anthropomorphized the pelicans, comparing the bird to a child in need of help. Her maternal response motivated her action and was tied to a deep spiritual belief that we are being judged by how we treat animals. Exploring the life story of interviewees can tell us a great deal about how our ability to feel compassion is created through life experience. The interview process can also be used to uncover what inhibits our capacity to feel. In the case of the wildlife agents, the use of words such as *resource* to describe the pelicans deadened their individual capacity for empathy. The emotional distance that this created, however, did enable them to focus on broader issues such as the protection of habitats.

We have a complicated and confused relationship to animal life. It is characterized by heartless neglect and flickering moments of compassion. Many of the interviewees in this project expressed a personal guilt for the oil spill since they drove a car and used gasoline, and they noted that guilt as one of their motivations for participating in the response. But when I asked them about the moratorium on offshore drilling that was imposed shortly after the spill, many of these same interviewees said that it was a bad idea because it was costing people their jobs and hurting the local economy. So even among those who had opened themselves up to the suffering that our abuse of the environment was causing, our obligations to animal life were still unclear. Our capacity for compassion and what triggers this compassion are things that are changing from generation to generation and from culture to culture. Understanding what constructs our ability to feel for and empathize with animal life is vital to comprehending and improving our relationship to nature.

With an emphasis on life story, oral history can provide us with important insights into why we feel, and it can reveal indicators of how our relationship with the natural world is evolving. It is clear that factors in our own life stories either nurture within us the ability to empathize with the other creatures that share our planet, or inhibit the development of that capacity. In the 1982 film *Blade Runner*, the hero, Rick Deckard, played by actor Harrison Ford, is a bounty hunter in search of rogue androids who are posing as human beings. To determine who is human, Deckard uses an interview process intended to provoke empathetic responses in interviewees. In the interview, he delves into the subject's life story and asks interviewees questions about how they would react to certain situations, often involving the suffering of animals. Their reaction or lack of reaction determined whether or not they were human.[9] Although the oral historian can be fairly certain that their narrators are in fact human, using interviews to explore their compassion for other living things can tell us a great deal about our own humanity. Our treatment of animals, and the environment which they inhabit, calls out for an "empathy test." Oral history can be an important part of this test.

Notes

1. Interview with Tom Benson by the author, May 28, 2013. Benson is the long-time owner of The New Orleans Saints football team. He purchased the local NBA franchise The New Orleans Hornets in 2012. During the oil spill, Benson and members of his New Orleans Saints visited the bird rehabilitation center for a media opportunity. The experience inspired Benson to rename the New Orleans Hornets the New Orleans Pelicans.
2. "History of Oil and Gas in Louisiana and the Gulf Coast Region," http://www.dnr.louisiana.gov/assets/TAD/education/BGBB/6/la_oil.html.
3. Tyler Priest, *The Offshore Imperative: Shell Oil's Search for Petroleum in Postwar America* (College Station: Texas A&M University Press, 2007), 1–10. See also Diane E. Austin, ed., *History of the Offshore Oil and Gas Industry in Southern Louisiana, Volume V: Guide to the Interviews* (New Orleans: US Department of the Interior, Minerals Management Service Gulf of Mexico OCS Region, 2008).
4. "Pelicans, Back from Brink of Extinction: Face Oil Threat," *New York Times*, June 5, 2010; United States Fish and Wildlife Service, "Brown Pelican Proposed Delisting Questions and Answers," https://www.nps.gov/chis/learn/news/upload/pelican%20delisting%20questions%20and%20answers.pdf; Coastal Wetlands Planning, Protection and Restoration Act, "The Brown Pelican's Return to Costal Louisiana," https://lacoast.gov/new/ed/articles/brownpelican.aspx
5. International Bird Rescue, "Pioneers in Oiled Aquatic Bird Care," https://www.birdrescue.org/about/history/
6. There are numerous media treatments regarding the limits that BP was imposing on media access to the spill; some include: Jeremy W. Peters, "Efforts to Limit the Flow of Spill News," *New York Times*, June 9, 2010; Matthew Philips, "Photographers Say BP Restricts Access to Oil Spill," *Newsweek*, May 25, 2010; Bridget Desimone, "In Reporting on Oil Spill, Limits Persist on Media Access in the Gulf," *PBS Newshour*, June 30, 2010.

7. Louisiana State Animal Response Team, https://www.lsart.org. LSART was organized with the assistance of the Louisiana State Veterinarians Office. In addition to providing BP with labor during the Deepwater Horizon Oil Spill, they have responded to animal issues caused by hurricanes Katrina, Gustav, and Ike.
8. President Theodore Roosevelt established the Breton National Wildlife Refuge in 1904. It is accessible only by boat. Breton Island was home to thousands of nesting pairs of brown pelicans, royal terns, Caspian terns, and seagulls. The island was designated as a Globally Important Bird Area, and the threat posed to the island by the spill triggered international attention.
9. *Blade Runner* was based on the science fiction novel *Do Androids Dream of Electric Sheep?* (Doubleday, 1968) by Philip K. Dick. The interview technique was called the Voigt-Kampff empathy test; an example of questions asked is, "You're in a desert walking along in the sand when all of a sudden you look down, and you see a tortoise, it's crawling toward you. You reach down, you flip the tortoise on its back. The tortoise lays on its back, its belly baking in the hot sun, beating its legs trying to turn itself over, but it can't, not without your help. But you're not helping. Why is that?"

3

Fragmentary Time

Memory and Politics in the Wake of the *Torrey Canyon*

Timothy Cooper and Anna Green

Oral history by Anna Green with environmental activist "Jane Evans" at St. Agnes, Cornwall, UK, June 2012[1]

In 1967, the oil tanker Torrey Canyon *went aground on the notorious Seven Stones reef fifteen miles off the Cornish coast, spilling most of its 119,000 tons of crude oil into the sea. Due to the prevailing winds and exceptionally high spring tides, the oil seriously contaminated both the coastline of Cornwall and the coast of Brittany. It became the world's first major oil tanker disaster. In 2012, we made a public call seeking those willing to record an oral history interview about their memories of the disaster. The request generated more than one hundred responses, and we were able to record the memories of fifty men and women. They included both residents living in villages along the Cornish coast and members of the army, navy, and fire service drafted in to clean up the oil.*

The interviews followed an open-ended format in three parts: autobiographical information and life narrative, specific memories of the events and impact of the Torrey Canyon, *and reflection upon the meaning of the oil spill in the longer term. Evans studied zoology at university, became a teacher, and has been involved in environmental causes throughout her life, including with the Cornwall Wildlife Trust and the Friends of the Earth. She was fourteen years old at the time of the disaster.*

The authors thank Sarah Bulmer for her comments on an early version of this chapter. The research for this chapter was undertaken with the support of British Academy/Leverhulme Trust Small Grant 110572.

Timothy Cooper and Anna Green, *Fragmentary Time* In: *Oral History and the Environment.* Edited by Stephen M. Sloan and Mark Cave, Oxford University Press. © Oxford University Press 2022. DOI: 10.1093/oso/9780190684969.003.0004

EVANS: My nickname was the "harbor rat," I spent a lot of time without shoes on and looking in rock pools and all that kind of thing. I was fortunate enough to go to grammar school, and when the *Torrey Canyon* happened in 1967 it was a horrible shock to see the beaches being polluted so badly and to see all the birds, it was just horrible seeing gannets, you know...

GREEN: You've got quite a clear visual memory of it?

EVANS: Most definitely, guillemots, razorbills, because I was actually involved with helping to pick them up off the beach, count the numbers, and there was also an RSPCA [Royal Society for the Prevention of Cruelty to Animals] center at Blowing house near Perranporth, can't remember the name of the RSPCA inspector... I remember going over to see some of them being rehabilitated and to help with the washing—*[you did?]*—Yeah, yeah. They had to tape their bills up because, obviously, they were quite fierce but some of them were in a terrible state, and of course if they ingested any of the oil trying to preen themselves, it's incredibly toxic. And I have very distinct memories, it was mostly around Fistral Beach, Fistral and Newquay where the oil was coming in as I remember, and obviously people were very worried about the tourist trade and so on, but I was more worried about the wildlife. But not everyone seemed to be very interested in it. My family were friends with a banking... a banker's wife and children, and I remember one of the teenage sons being very sniffy about me, sort of suggesting "you write to the pope and ask for a medal" or something. Really, not being sympathetic *[laughter]*. But it did have quite a long-lasting effect because they sprayed, they bombed the ship didn't they, and then they sprayed with a lot of detergent to try and either sink it or disperse it, I don't know, and I think that was toxic to a lot of the mollusks; the grazing mollusks like limpets. So their numbers plummeted as I recall, and then because they weren't grazing, the grazing pressure had gone, I remember the rocks around Newquay, around the island on Towan Beach and so on, was absolutely bright green in the following year because all the algae, I suppose it was either *Ulva lactuca* or *Enteromorpha intestinalis*, came back with a vengeance and then gradually the mollusk numbers recovered and I suppose the limpets had a pretty good time, the ecology was definitely upset and around the island there were quite deep rock pools and you would get things like snakeslock anemones and they disappeared quite some time ago, so whether the toxic effects of that or just the disturbance, too many people poking around the rock pools. Having said that, I don't know.

GREEN: They didn't come back?

EVANS: They're not around, just in front of where the aquarium is now, I haven't seen them round there for years and years, so that's a shame. But, it was a formative time, I suppose, because I was always interested in natural history, and it encouraged me to go on and read zoology. I was at grammar school and they only taught zoology, they didn't teach biology or botany. But the teacher was Tom Burley, who was a very inspirational teacher, and he's better known as the person who wrote the books that made the Wycliffe detective series. He had an interesting background. He was I think trained as an engineer originally, went to work for the gas board or whatever they were called. Didn't like it, went back to Oxford and retrained in zoology and came into teaching. And he was very thought-provoking and stimulating, and he took myself and one of the children of the then County Secretary for Education I think it was, all of whose children did fantastic things . . . and another chap who became a vet, and he took us under his wing a bit and coached us a bit privately. I don't think . . . I'm the only one that didn't really live up to expectations perhaps, but I went to university to read zoology.

GREEN: Where did you go?

EVANS: I went to Cardiff, in South Wales. I didn't want to be too far north, and with a surname like [mine] in the 70s, they took you without too much—

GREEN: —Assumed you were Welsh?

EVANS: Exactly. They didn't even interview, I think. So yes, it was an interesting time to live through. First time I had any inkling about environmental degradation and how bad pollution could be. And that encouraged me then to be involved with Friends of the Earth when it first started when I was sixteen, and I've been involved on and off ever since. I became a water pollution activist for a while.

GREEN: So you really see the *Torrey Canyon* as very important in your own . . .

EVANS: I suppose it was, I hadn't really thought about it, but reading the article in the West Briton and asking for anyone with memories to come forward I thought, well, you know, I do remember it and I've always been involved with environmental stuff. So I've been supporting the Cornwall Wildlife Trust, not quite since it started, it was just a group of naturalists. I've been involved for, I suppose, about thirty years or something like that. My daughter was very little when I got involved, and I'm currently vice president and sit on the council. Yeah.

GREEN: What were you involved with in terms of water pollution?

EVANS: It was to do with the coast around here, before Surfers Against Sewage started, the Friends of the Earth group—Perranporth and St. Agnes Friends of the Earth—we'd been involved in recycling, we used to actually go 'round and collect people's paper and you know, recycle it. It's quite hard physical work. We had demonstrations on the beach. We even had a plane fly-past towing a banner, warning people about the dangers of sewage pollution, and we could never really understand why hoteliers and businesses would be upset, because if you could have a decent system treating sewage like they have in Jersey, the water coming out of the effluent pipe is cleaner than the sea water, you could bathe more safely in that. Surely that would be a fantastic sort of advert for the place. So we did that for a while, and we were . . . my friend Sarah W., put together reports every year on the bathing water quality, and analyzed stuff, and then as we got older I suppose we devolved into a quilting group, as we got married and what have you, and then, we've now devolved into making our daughters quilts *[laughter]*, cot quilts and this kind of thing, and because we all tend to be vegetarian or vegan, catering for the weddings as well, because it was quite hard to get decent food.

GREEN: There's a group of you for quite a long time?

EVANS: Just a group . . .

GREEN: And ones you knew when you were growing up?

EVANS: No . . .

GREEN: Or did you meet later, through Friends of the Earth?

EVANS: Yes, through Friends of the Earth, yes. Now Surfers Against Sewage is doing that sort of job, you know, there's no need for us to continue with that, so that's good. But I'm currently interested in the transition town idea, and St. Agnes has a group, and the partner of one of the people who's been in the Friends of the Earth group and the quilting group and so on, he's chairman of that group and I'm doing more voluntary work for the Wildlife Trust since I retired as a teacher. So there's a voluntary marine conservation area in the village, and I'm involved with that and running events. And I'm a licensed bat warden as well, so coming up in July there's a meeting celebrating the coast where we're moth-trapping and looking for bats on the coast.

GREEN: So, your interests sort of encompass both the coast and the marine life, and . . .

EVANS: Yes, well more, not so much the marine life since I've had a family and what have you, but I've always been interested in everything, so *[pause]*

I've been very lucky to teach biology because it's sort of my hobby as well as my job, you know. I was just very fortunate.

Here, Evans speaks about seeing the disaster on television and the role of the RSPCA in the clean-up.

GREEN: Have you taken notice of other oil spills in other parts of the world?

EVANS: Well, *Exxon Valdez*, I mean, that was horrendous in that pristine environment, but also when it's so much colder the bacteria and the microbes work so much more slowly, although it is an organic material it's going to take much, much longer to disappear. I do have a recollection of the rocks being stained black and I suppose that has dissipated over time, partly wave action, bleaching of the sun, but it must be to do with microbes as well I think.

GREEN: In terms of, perhaps thinking of Cornwall, and thinking of the *Torrey Canyon* in terms of Cornish history or environmental history, how would you see the *Torrey Canyon*? How would you locate it, I suppose I'm trying to say, in terms of significance?

EVANS: Well, it has played a significant part because the Cornwall Wildlife Trust have tried to push this quite a bit, because we have such interesting environments, like the Falmouth area, with the ria.[2] Pollution is obviously one problem, protecting the environment is obviously a big issue at the moment about dredging the very rare maerl beds, and deepening the harbor so that they can get much bigger ships in.[3] So yes, I think it has played a significant part. But the trouble is, I suppose, one disaster happens and people just plough on, and another generation has forgotten about it, and it fades in the memory. But at least people have these booms and rapid response means now, which they didn't have in those days I suppose. It could have been contained perhaps much better than it was, I don't know. With the rough seas perhaps that wouldn't have been possible, even if they had the equipment.

GREEN: I think you said earlier it was the first one you were aware of, in terms of environmental disaster of that kind.

EVANS: Well, the thing was we didn't, my mother listened to Radio One, and we watched the national news and that was really all the media that there were coming into the house. So, I wouldn't have been aware of other things going on, perhaps, I think I was very sheltered really, until I went to university.

GREEN: Do you have any sense that it played a role in environmentalism in Cornwall? Among your generation, or . . .

EVANS: I haven't heard it spoken of.

GREEN: You haven't, no.

EVANS: No, I haven't.

GREEN: Not mentioned as a . . .

EVANS: No, as a seminal sort of thing, no, I can't say so.

GREEN: What would you see as being sort of more significant issues, if you like?

EVANS: Sewage pollution in the sea, and I suppose Surfers Against Sewage really captured people's imaginations, but actually the Friends of the Earth group, we'd been doing a lot of this, and in fact before they had their own technical people we were doing that side of things for them. We didn't care about the publicity because as long as there was a job being done by somebody, that was fine.

And I suppose habitat destruction generally, hedgerow removal was another big sort of thing that I remember. And now your concept is of joining up contiguous areas to make living landscapes and also to give wildlife corridors, so that as the climate change happens things can migrate north or wherever they need to along wildlife corridors. Those I would say are some of the bigger things and, you know, we've only recently, relatively recently at the Wildlife Trust had a marine conservation officer. Now there is a significant marine team, and it's long overdue. But with a charity you can only move at a certain pace, you know, you can't expand and then have panics and have to shed people and so on. So, it's had to be very slow development. And there is a Marine Bill that I suppose is a big thing, that has gone through, is going through, hopefully. But I must admit, I tend to sort of be amazed at how resilient the environment appears to be. But I think it has already been degraded a great deal. My head of department at Truro School when I first went there was a Cornishman, Dr. Blake, and he remembers as a boy in the Polzeath area playing a game where you tried to get across the beach without standing on any lugworm casts. Well you would be lucky to find one these days.

GREEN: Really.

EVANS: Yes, you tend still to think it's an abundant environment, but actually I think we've lost an awful lot and we've got no baseline to measure against, really. We're only just now starting to get the baselines recorded.

GREEN: What is, and this is a very general question, your perception of the attitude towards the environment in Cornwall? You're at the front line of it, you've been fighting for things . . .

EVANS: I'm very disappointed, as a teacher, because young kids are very into environmental things, and then as they get older, I don't know, perhaps it just doesn't seem very sexy going out with a magnifying glass and poking around in rock pools, that sort of thing. And certainly, I think a lot of teenagers sort of feel a bit defeated by the scale of the problems and so they perhaps think "I'll just enjoy my life and get what I can out of it and blow the next generation." That's the sort of attitude I've come across as a teacher, and I've met very, very few pupils who have ever gone into ecology conservation work. Most of them have gone into professions which pay pretty well, like industry, law, medicine, pet science, that kind of thing.

Commentary

The loss of the *Torrey Canyon* off the coast of Cornwall in March 1967, and the subsequent month-long clean-up operation, has often been regarded as one of the turning points in the making of popular environmental consciousness in Britain.[4] John Sheail called the incident probably "the most shocking man-induced disaster in peacetime," and while this might be somewhat hyperbolic, there are certainly good reasons for attributing historic significance to the *Torrey Canyon* disaster.[5] For one thing, the oil spill occurred in the middle of a decade that saw the publication of Rachel Carson's *Silent Spring* in 1962, and ended with the first "Earth Day" in 1970—ultimately culminating in the United Nations Conference on the Human Environment in 1972.[6] The spill and clean-up operation also received very widespread media coverage, which meant that it contributed to deepening the emerging language, and particularly the visual tropes, of catastrophist environmentalism.[7] Certainly, an environmentalist like John Lister-Kaye later looked back at the experience of the disaster as a damascene moment in his life story.[8] The disaster, and similar events that followed, also left an enduring presence in popular culture in novels like Ralph Hammond Innes' *Black Tide*.[9] Indeed, down to the present day the British media produce regular decennial anniversary commentaries that construct an "official" line on the spill, in which it was instrumental in the formation of public and political interest in environmental issues.[10] All these are stories that "make sense" of the catastrophe, partly redeeming it through its role in heightening contemporary awareness of environmental issues.

Seen from these perspectives, the *Torrey Canyon* has an apparently unassailable claim to be a key event in the rise of modern popular concern with environmental issues. Yet, Evans's account above appears to offer a rather different perspective. Green interviewed Evans in the summer of 2012. She was one of almost a hundred people who had responded to a public call for participants in an oral history project on the disaster.[11] We interviewed nearly fifty of them. While most of these narrators had firmly established stories of the disaster and its place in their lives, they did not generally express the idea that the event caused them to develop an explicitly green consciousness, nor were most engaged in any form of environmental activism. The fact that so many people should remember this event yet not recall it as the source of personal environmental awareness is suggestive that, in Cornish memory at least, this event has a rather different meaning to that presented by public and historical memory. Evans's story is, therefore, exceptional. Is it the one

Surrounded by a thick swirl of oil, firemen and fisherman survey the results of the SS *Torrey Canyon* oil spill from a rowboat in Porthleven, Cornwall, March 29, 1967. *PA Images / Alamy Stock Photo*

instance of a clear connection between environmental disaster and later consciousness and activism?

Here, we suggest that despite initial appearances, Evans's account also resists integration of the disaster into a linear narrative of the "making of modern environmentalism." As oral historians have shown in other contexts, there is often dissonance between the official stories of important or traumatic events and actual community memory.[12] Indeed, previous studies of the impact of the *Torrey Canyon* have largely focused on the official response and public policy consequences, revealing the influence of the catastrophe on environmental governance and politics in the United Kingdom.[13] On the other hand, community and social memory of the event has been almost entirely neglected. Evans's account of the disaster, and its relationship to her life story, implicitly deconstructs this historical appropriation of the disaster. Read against the grain of the established narratives of the *Torrey Canyon*, her account reveals a set of reactions that are much more fragmentary in form. Evans's account, which eschews linear storytelling, reveals the danger of drawing overly confident connections between environmental disaster and the transformation of everyday consciousness. Instead, it suggests an ongoing uneasy and uneven relationship between the lived experience of the catastrophe and environmental activism.

Form is an important part of the fragmentary character of Evans's story. In fact, and in contrast to many others that we interviewed, Evans chooses not to use a straightforward linear historical narrative in her telling. Nor does she focus only on the well-known moments of the event, such as the bombing of the ship, which are common elements in other accounts. Rather, Evans seems to reject historical narrative in favor of a more episodic, discontinuous form. She weaves memories of the *Torrey Canyon* incident itself with recollections of her subsequent work in conservationism into a complex stream-of-consciousness account. Throughout the interview she repeatedly moves between points in time. Memories of the incident itself overlap with those of later activism or personal educational history. Consequently, events from different moments in Evans's life story appear as juxtaposed with the catastrophe, and the listener is asked to make connections across time and space.

Listening again to this interview, the effect of this use of non-linear narrative, the creation of a sense of fragmentary time, appears more than contingent. It immediately transforms the significance accorded to the disaster for a listener, decentering it in their own mind, implicitly asking them to question the significance of the event itself. For researchers coming to this project presuming the significance of *Torrey Canyon* in some way, the effect is palpable and a little disturbing. It forces a reconsideration of the significance of the project we were undertaking. Here we encounter the events of 1967 as just one passing moment in a lifetime of commitment to both the study and conservation of nature. Evans does not even present the *Torrey Canyon* disaster as *the* most significant event in a developing passion and concern for nature. Her fragmentary form of narration ensures that we are not permitted to assume a single point of origin from which to make sense of this story. Throughout the interview Evans persistently "resists" coming back to the issue of the *Torrey Canyon*, and Green works hard throughout the interview to gently return to it.

One might read this refusal to straightforwardly affirm the historical significance of the *Torrey Canyon* as simply a contingent part of the interview process: a reflection of the differing objectives of the storyteller and the interviewer/listener. However, it may also be that there is more at work here in Evans's choice of form and presentation of her story. For instance, perhaps Evans's mode of storytelling can be regarded as a rejection of the "masculine" form of historical narration with its reliance on making events cohere around causes and linear time.[14] Sociologist Niamh Moore has observed how women's activist life narratives often have to challenge the marginalization of women's experiences in environmental campaigning and commonly have "an entirely different aim from conventional history."[15] From this interpretive perspective, perhaps Evans's use of fragmented narrative can be read as resort to a form that reflects the experience of women's stories of environmentalist commitment more generally. Certainly, Evans's determined refusal of the historical narrative form mitigates against the kind of account of the *Torrey Canyon* as a "conversion" experience on the road to enlightenment of the type given in autobiographical accounts like Lister-Kaye's. In lived experience, there are rarely such neat connections between events—even traumatic events—and individual political commitments. This is something that historical accounts of the origins of modern environmentalism, which are largely based on written sources, often miss. Precisely because one of the effects of bureaucratic archives is to efface the complexity of real events as they are experienced through a lifetime, they render them more intelligible and coherent than they maybe in everyday life. Evans's account reveals the disjointed, complex temporality shaping everyday responses to disasters.[16]

Following on from Moore's observation of the importance of gendered contexts in women's environmental activism, it is similarly apparent how the course of Evans's life has been mediated through gendered social structures, especially the family. Evans returns to the issue of family life several times throughout her interview. She mentions that "her family" were friends with a banker's family, signaling early the significance of this level of experience. Later, family responsibilities had an important role in modulating environmental activism for Evans, significantly affecting the scope for her activism. Evans's collaboration with a Friends of the Earth group changed, we are told, "As we got married and what have you, and then, we've devolved into making our daughters quilts *[laughter]*, cot quilts and this kind of thing, and because we all tend to be vegetarian or vegan, catering for the weddings as well, because it was quite hard to get decent food." Clearly, Evans's involvement with environmental causes has moved between formal and informal contexts over her life course. She marks this as a "devolution," as if quilting or vegetarian catering are a move away from "proper" politics. However, she also shows us the intertwining of her environmental concerns with private life, in which a daughter's wedding becomes a point for exploring and extending a commitment to vegetarianism. Equally, Evans's interest in marine coastal protection is revealed as affected by having a family and the new commitments that brings. The enormous value of this account is that it shows how Evans's environmental activism did not emerge smoothly from the experience of a single transformative event but was compelled to evolve in tandem with the changing demands of everyday life for a woman in twentieth-century Britain. Evans's concern for nature has always found a consistent outlet, but the avenues for this expression have been different at different times in life.

The fragmentary structure of Evans's story also hints at a problem faced by narrators in telling a story about a highly mediated event, widely covered at the time in the national and global newspapers, on television, and on radio. Evans's memories are clearly a compound of different sources as is evident from the rest of the interview but not transcribed above. Some are a matter of personal direct experience, as in collecting oiled birds and borrowing gloves and wellies for this. Then there are stories told by others, perhaps overheard, and remembered because of a powerful image or humorous content. For instance, the friend of the parents whose new car's paint was damaged by the detergent used to disperse the oil on the sea. Then there are memories of media images, the bombing of the ship, for example. All these various points of reference are suggestive of the nature of memory in this case, composed of an ensemble of images from different sources, and not necessarily easily narratable through a historical or autobiographical mode. An event remembered both through the media and personal experience struggles to be disentangled. It all suggests that the experience of the modern environmental disaster was, for Evans, partly also a mediated experience and thereby a fragmentary one, an event that was topographically distant and proximate in the same moment.

The fragmentary nature of Evans's story continues beyond form and into content. Evans begins by establishing her presence and rootedness within the community. She recalls being nicknamed "harbor rat," placing herself affectionately within the everyday life of the village. Yet, Evans also knows that she observed that world from a relatively privileged position compared to some of her neighbors. There is a sense in her story of simultaneous presence and distance between her family and the wider community. This capacity to distance herself from her immediate surroundings is revealed when she places the meaning of the events of 1967 into a different framework from that of the wider community: "I remember," she says, "obviously, people were very worried about the tourist trade and so on, but I was more worried about the wildlife. But not everyone seemed to be very interested in it." She repeats this sense of difference at various points throughout her telling. One is her reference to a lifelong commitment to vegetarianism, mentioned several times during the interview. Another is when she recounts the story about an unsympathetic friend's response to her work collecting oiled seabirds. These moments of what was clearly a strongly felt difference from those around her emphasize that Evans's story is being told simultaneously from within and without the wider Cornish community. Evans is a part of this world and understands its fears and motivations, but she also emphasizes that she has always had a different set of values.

Perhaps the strongest emphasis on difference is apparent in Evans's account of her educational experience. It is here that we get a sense of why, perhaps, the *Torrey Canyon* incident may have had such a strong effect. Indeed, Evans remembers the entire period around 1967 as being a "formative" one, particularly in terms of her education. She recalls being "fortunate enough" to go to grammar school, and clearly remembers her school days fondly. She emphasizes the influence of her zoology teacher in interesting her in the subject she later went on to study at university, an encounter that ultimately led to a career teaching biology. Throughout the interview Evans reveals her professional expert knowledge. Evans's use of binomial names for local flora and fauna sets up a significant marker of educational difference. This clearly places Evans in a unique position from which to assess the significance of events, as simultaneously an insider (a member of the community) and an outsider (one of the learned, scientific, community).

Evans's possession of expert knowledge does not, however, obscure an effective response. Both personal horror and professional curiosity combine in Evans's combined remembering of disasters such as the *Exxon Valdez* spill, which "was horrendous in that pristine environment" although, we are simultaneously reminded, "when it's so much colder the bacteria and the microbes work so much more slowly, although it is an organic material it's going to take much, much longer to disappear." Here she combines aesthetic and affective reactions with concerns drawn from her scientific understanding of the comparative ecological effects of oil in different environments. Given the importance accorded to the role of a solidifying ecological science in many accounts of modern of environmentalism, one is prompted to ask where precisely "environmental consciousness" originates for Evans, whether in the emotional, embodied response to experience or in the realm of scientific concern. Of course, the borders between these epistemologies are always hazy and perhaps even unidentifiable.

Throughout her story, Evans presents her commitment to environmental activism as the consequence of an unexpected unraveling of moments. It has been a lifelong process of commitment and curiosity rather than a sudden conversion to a cause. Her account of life as an activist is told in the context of the *Torrey Canyon*, but the significance of the disaster to her activism shifts dramatically throughout the telling. On the one hand, she suggests that the incident was the first time that she had a clear insight into the significance of pollution, and that it was one of the factors encouraging her early involvement with the nascent Friends of the Earth organization.[17] On the other hand, and simultaneously, she reminds us that she had not necessarily made this connection before her involvement in this oral history project. Evans even distances herself from the narrative details of the disaster. "I know nothing about its origin," she states, "who owned it, where it came from, where it was going to." Indeed, given the chance by Green to explicitly place *Torrey Canyon* into a narrative of environmental awakening, Evans explicitly rejects it, replying that "I haven't heard it spoken of."

Each time the question arises of the connection between the *Torrey Canyon* and her activism, Evans equivocates. In the context of her life story, it might be that Evans's interests in nature, and her commitment to vegetarianism,

On a beach near Hayle, Cornwall, a bulldozer works to remove oil washed up from the *Torrey Canyon* spill, May 1, 1967. *PA Images / Alamy Stock Photo*

would very well have led her in the same direction. These are, after all, deep personal commitments that have been a part of the substance of her daily life, whereas the events of the *Torrey Canyon* were a passing, if intense, disruption. Throughout the interview, Evans is keen to emphasize her *lifelong* participation in nature conservation and environmental projects. It is a rich story that intersects with a diverse number British environmental and conservation organizations. She places her membership of the Cornwall Wildlife Trust, a naturalist organization that runs several local reserves, alongside her involvement in the politics of sewage pollution. She recounts her involvement with paper recycling with Friends of the Earth, an activity common among grass-roots environmentalists in the 1970s and 1980s.[18] This account of a long activist life, however, raises its own questions about the role of these organizations themselves in sustaining local memory of the *Torrey Canyon*. Evans's membership of the Wildlife Trust would suggest access to an institutional memory of the disaster. Indeed, the Trust's predecessor, the Cornwall Naturalists' Trust, strongly made the conservation case to the government during the catastrophe, expressing concerns regarding use of chemical oil dispersants.[19] Evans does not at any point recall this connection. Perhaps this involvement even been forgotten within the organization itself? If so, what significance can we attribute to the events of 1967 in making local conservation and environmentalist sentiment?

Evans's own activism has moved through a series of phases. She helped produce reports for Friends of the Earth on water quality around the Cornwall coast during the 1990s, part of a campaign to address sewage pollution in the area.[20] She observes that this was an important aspect of the work of the Friends of the Earth group in St Agnes "before Surfers Against Sewage started," revealing how her political commitments must be seen in a changing wider context. One does not just *become* an environmentalist, it appears, rather the impulse to support different environmental causes is present over the context of a life course. Evans observes how it was that the same women active in water pollution campaigns also formed part of the "quilting group," which she so self-deprecatingly describes in zoological terms as a "devolution." This devaluing of a creative practice relative to "proper" political activity suggests a kind of hierarchy of action, but we should not necessarily accept Evans's self-characterization of this activity. Quilting and campaigning for clean seas are part of a larger way of living for Evans; along with her vegetarianism, they are markers of identity and difference. Her story may prompt us to ask where environmental activism begins or ends—although Evans privileges formal activism as, in some sense, more truly *political*, she also shows that in everyday life the boundaries between forms of environmentalist self-expression are blurred. As she points out, one of the members of the quilting group later also became chair of the local transition town initiative.

Intriguingly, Evans delves deeper into the uncertainties of historical causation as she discusses ecological knowledge and the impact of oil and other pollution around the Cornish coast. This encounter with material nature again troubles and fragments historical time, which proves inadequate to narrating the story of environmental change. Evans discusses the effects of the chemical dispersant used as part of the clean-up operation, but couches her memory of this in extremely conditional and ambivalent terms. At this point, in her account the retelling brings together different instants in time, and different causes. The effects of spraying dispersants were followed by algal blooms the following year and the death of "grazing mollusks." Then Evans's tells an account of the "snakeslock anemones" disappearing, as she puts it, "quite some time ago." She speculates that perhaps this was a result of toxic detergent, but then adds that it may also be a consequence of the feet of countless tourists over the years. Here, Evans's narrative encounters time and duration as signifiers of ambiguity breaking the discrete event down into something more uncertain, accretive, and fragmentary. Finally, Evans suddenly moves this story back into the ecological present: "They're not around in front of where the aquarium is now." This juxtaposition of different moments in time demonstrates the extreme difficulty of attributing ecological change with certainty to any one cause. This is not the result of ignorance. Quite the opposite. It is Evans's ecological knowledge that allows her access to a complex and uncertain picture; to comprehend the possible connection between a changing social and economic context and local ecologies. In making this connection and drawing our attention to a wider span of time, Evans points to an untold story beyond the *Torrey Canyon* disaster, a tale of a slower, possibly more devastating violence caused by the effects of tourism on Cornwall's natural environment.[21]

At the end of the interview, Evans's reflections on the longer-term implications of the *Torrey Canyon* shift toward contemporary environmental concerns. She refers to present-day worries over proposals to dredge Falmouth Harbour. The opposition of local activists to dredging has been counterpointed by arguments about the economic necessity of retaining the docks and the cruise ship trade.[22] Evans observes that, "one disaster happens, and people just plough on, and another generation has forgotten about it, and it fades in the memory." Just as it is difficult to pinpoint 1967 as a point of origin for the concerns of someone like Evans, so its significance to the wider community is also in doubt. Evans is pessimistic about the apparent lack of interest among contemporary young people for environmental causes. She is "disappointed" that early years' interest in nature seems to die out as young people enter their late teens—although we might recall here the scepticism of Evans's own youthful peers when confronted with her concern for local birdlife. We do not have to accept this account of contemporary youth to see at work a troubled uncertainty about the environmentalist's place in the wider community.

Environmental historians have long probed the question of the origins of modern environmentalism.[23] The histories of conservationism and environmentalism as ideas among elites have become a central part of accounts of environmental governance, regarded by some as expressions of the "paternalist ideology of command."[24] This is undoubtedly an important element of environmental "consciousness." However, the larger question of how environmental ideas became part of *popular* political culture remains woefully under-explored.[25] How far future explorations like this may allow us to fundamentally rethink the story of modern environmentalism remains to be explored. However, it is clear that Evans's own story of a life of environmental activism at least raises a question about overly neat accounts of the "rise" or "making" of environmental consciousness. There was no neat connection between Evans's activist life and her experience of environmental disaster. Rather, the *Torrey Canyon* was one part of a varied life experience in which oiled birds loomed larger or smaller depending on the moment.

The search for the origins of modern environmentalism is founded in a conception of history dependent upon linear time. From this perspective, the *Torrey Canyon* matters because it is an iterative event leading to a modern environmental consciousness. But disasters like the *Torrey Canyon* should not simply be regarded as one part of a chain of "causes" generating traceable shifts in public understandings of nature. Rather, they are complex "moments," simultaneously embedded in and disruptive of the wider rhythms of everyday life.[26] Even for an avowed environmental campaigner like Evans, it is difficult to integrate these moments into a straightforward story of growing environmental awareness. This raises the question of what a history of environmentalism might look like beyond a search for origins.[27] Recently, historians, reflecting on the political disappointments that Evans recounts, have suggested that the real significance of environmentalism was not the way it changed our conceptions of nature but rather what it left unchanged.[28] Perhaps, then, historians, need to engage more fully not with a grand narrative of the making of modern environmentalism, but with its relative marginality to contemporary society as a whole. To do this, oral histories that reveal the complexities and ambiguities of popular environmental thought will be essential. They, and the texture of everyday experience that they reveal, are a critical missing component in our understanding of the fragmentary and unfinished process of greening the modern world.

Notes

1. A pseudonym is used for the interviewee in excerpts presented here.
2. A ria is a funnel-shaped valley near a coastline that has been partially submerged by water and has a branch-like appearance. There are a number of these along Cornwall's "drowned coastline"; this is most likely a reference to the famous Helford River area south of Falmouth Bay in Cornwall.
3. Maerl is a purple-pink hard seaweed that provides important habitat for a variety of marine life.
4. David Peterson Del Mar, *Environmentalism* (New York: Routledge, 2006), 103.
5. John Sheail, "Torrey Canyon: The Political Dimension," *Journal of Contemporary History* 42, no. 3 (1 July 2007): 221.
6. Del Mar, *Environmentalism*; Thorsten Schulz-Walden, *Anfänge globaler Umweltpolitik: Umweltsicherheit in der internationalen Politik (1969–1975)* (Munich: Walter de Gruyter, 2013).
7. Kathryn Morse, "There Will Be Birds: Images of Oil Disasters in the Nineteenth and Twentieth Centuries," *Journal of American History* 99, no. 1 (June 1, 2012): 124–34.
8. John Lister-Kaye, *Song of the Rolling Earth: A Highland Odyssey* (London: Little, Brown, 2004).
9. Hammond Innes, *The Black Tide* (London: Random House, 2013).
10. Adam Vaughan, "Torrey Canyon Disaster – the UK's Worst-Ever Oil Spill 50 Years On," *The Guardian*, March 18, 2017.
11. Further information about the research project may be found in Anna Green and Timothy Cooper, "Community and Exclusion: The *Torrey Canyon* Disaster of 1967," *Journal of Social History* 48, no. 4 (2015): 892–909.
12. Alessandro Portelli, *The Death of Luigi Trastulli and Other Stories: Form and Meaning in Oral History* (Albany, NY: State University of New York Press, 1990); Suroopa Mukherjee, *Surviving Bhopal: Dancing Bodies, Written Texts, and Oral Testimonials of Women in the Wake of an Industrial Disaster* (Basingstoke: Palgrave Macmillan, 2010).
13. Sheail, "Torrey Canyon."
14. Elizabeth Grosz, *Time Travels: Feminism, Nature, Power*, (Durham, NC: Duke University Press, 2005).
15. Niamh Moore, "Paths to Ecofeminist Activism: Life-stories from the north-east of England," in *The Roots of Environmental Consciousness*, ed. Stephen Hussey and Paul Thompson (London: Routledge, 2005), 160–173, this quote is page 166.
16. Timothy Cooper, "Risk, Time and Everyday Environmentalism in Modern Britain," in., *Governing Risks in Modern Britain: Danger, Safety and Accidents, C. 1800–2000*, ed. Tom Crook and Mike Esbester (Basingstoke: Palgrave Macmillan, 2016), 149–68.
17. The United Kingdom arm of Friends of the Earth was formed in 1970.
18. Timothy Cooper, "War on Waste? The Politics of Waste and Recycling in Post-War Britain, 1950–1975," *Capitalism Nature Socialism* 20, no. 4 (December 2009): 53–72.
19. Cornwall Record Office AD 262/1, *Cornwall Naturalists Trust Bulletin*, July 1967.
20. *Bathing Water Quality in Cornwall* (Perranporth: Perranporth and St. Agnes Friends of the Earth, 1994).

21. Rob Nixon, *Slow Violence and the Environmentalism of the Poor* (Cambridge, MA: Harvard University Press, 2011).
22. "Falmouth Harbour Dredging Will Not Affect Wildlife, Study Shows," *BBC News*, http://www.bbc.co.uk/news/uk-england-cornwall-26756354.
23. Meredith Veldman, *Fantasy, The Bomb and the Greening of Britain: Romantic Protest, 1945–1980* (Cambridge: Cambridge University Press, 1994); Richard H. Grove, *Green Imperialism: Colonial Expansion, Tropical Island Edens and the Origins of Environmentalism, 1600–1860* (Cambridge: Cambridge University Press, 1995); Harriet Ritvo, *The Dawn of Green: Manchester, Thirlmere, and Modern Environmentalism* (London: University of Chicago Press, 2009); Fredrik Albritton Jonsson, *Enlightenment's Frontier: The Scottish Highlands and the Origins of Environmentalism*, The Lewis Walpole Series in Eighteenth-Century Culture and History (New Haven, CT: Yale University Press, 2013).
24. Richard Drayton, *Nature's Government: Science, Imperial Britain, and the 'Improvement' of the World* (New Haven, CT: Yale University Press, 2000).
25. Marco Armiero and Lise Sedrez, eds., *A History of Environmentalism: Local Struggles, Global Histories* (London: Bloomsbury Academic, 2014); J. F. M. Clark, "Pesticides, Pollution and the UK's Silent Spring, 1963–64: Poison in the Garden of England," *Notes and Records: The Royal Society Journal of the History of Science*, February 15, 2017, 20160040, https://doi.org/10.1098/rsnr.2016.0040.
26. Henri Lefebvre, *Rhythmanalysis: Space, Time, and Everyday Life* (London: Continuum, 2004).
27. Timothy Cooper and Anna Green, "The Torrey Canyon Disaster, Everyday Life, and the 'Greening' of Britain," *Environmental History* 22, no. 1 (1 January 2017): 101–26.
28. Jean Baptiste-Fressoz, "The Lessons of Disasters – Books & Ideas," https://booksandideas.net/The-Lessons-of-Disasters.html.

4

The Ghosts of Bhopal

Oral History, Environmental Justice, and the Literature of Protest

Suroopa Mukherjee

Oral histories by Bhopal Survivors' Movement Study with Bhopal activists Hazra Bi, Hamida Bi, Rashida Bi, Abdul Jabbar, and Satinath Sarangi, Bhopal, India, 2007, 2008, and 2012[1]

On December 3, 1984, Bhopal, the capital of Madhya Pradesh, India, was witness to an industrial disaster of such magnitude that its impact has been compared to Hiroshima.[2] *It was past midnight when forty tons of methyl isocyanate (MIC) leaked from inside the premises of Union Carbide India Limited, the Indian subsidiary of the multinational Union Carbide Corporation, killing or maiming anyone who came in its path. Decades later, the impact of the gas leak continues, a toxic legacy inherited by the residents of Bhopal.*

A lesser known fact about the tragedy is the environmental contamination that predated the disaster. A faulty disposal system spread hazardous waste indiscriminately during routine operation of the factory. This defect led to widespread ecological contamination that had a severe impact on the lives of people long before December 1984.[3]

What is least known about the Bhopal tragedy, however, is that India's oldest people's movement for environmental justice had its inception on the morning after the disaster and that this movement continues to this very day. Oral history was widely used by these activists to enable the affected community to present their own history in their own words, and to use these stories to create social change.[4]

Interviews with Hazra Bi, JP Nagar, December 12 and 16, 2007

A member of the International Campaign for Justice in Bhopal, Hazra Bi was interviewed by the Bhopal Survivors' Movement Study at the Sambhavna Trust Clinic.

Suroopa Mukherjee, *The Ghosts of Bhopal* In: *Oral History and the Environment.* Edited by Stephen M. Sloan and Mark Cave, Oxford University Press. © Oxford University Press 2022. DOI: 10.1093/oso/9780190684969.003.0005

BHOPAL SURVIVORS' MOVEMENT STUDY (BSMS): Has any incident changed your perspective on how to fight for your rights?

HAZRA BI: My perspective has changed. My son Mansoor never got proper medical care. I got a lot of motivation from the courts. When my son was summoned for the final settlement I was asked, "Are you Mansoor Ali's mother, your son gets Rs. 25,000, do you accept it?" I told the judge, "I do not want the money or this boy. I want my four-year-old son as he was before the gas leak."

The judge stared at me and asked, "What do you expect?" I said, "A doctor's report." One month later the doctors were called. Mansoor had a big file by that time, with eleven x-ray reports in it, because the hospital was his home. My son's childhood was lost. He could not read, write, or play. He was bedridden all the time with a saline drip attached to him. How far can a mother take this? This is not the question of me, Hajra Bi's son, but a lot of other mothers who went through the same thing.

BSMS: Can you tell us what was happening around you?

HAZRA BI: Moolchand's son Kishore also went through the same suffering, he finally died at the age of twenty-two. My neighbor Suman lost her eighteen-year-old son. There are many orphans in these colonies who are raised on other people's charity and leftovers. I have seen this with my own eyes. So that day in the courtroom I really got mad at the judge. He called me, read through the medical reports, and in the judgment Mansoor was declared normal.

BSMS: How did you hold on to your hope for justice?

HAZRA BI: I have a question for everyone. If my son is normal then why is he not playing and enjoying his life? Why does he remain admitted in hospital most of the time? My file was transferred to the head office, I received Rs. 55,000 from there. Then I got some hope. It is only through fighting that a person can claim his/her rights. To remain inactive will not fetch anything.

Interviews with Hamida Bi, Ibahim Pura, March 19 and April 27, 2008

This interview took place in the Swabhiman Kendra (Centre for Economic Self Reliance), which she headed. Hamida Bi was also founding member of the Bhopal Gas Peedit Mahila Udyog Sangathan (Bhopal Gas Affected Women Workers Union).

BSMS: What do you remember about third December?

HAMIDA BI: How can I forget? My daughter and son-in-law lived two streets away from me and they were missing. I went out to look for them. I had to go to the morgue, and there I witnessed many dead bodies being brought in trucks and vehicles and piled on the floor. I could not find my daughter there. Then I went to the Hamidia hospital, and the whole hospital was empty because the doctors had advised the patients to flee because the gas victims would soon flood the hospital. I was told my granddaughter was also missing. I found her near the water tank, someone had wrapped her in a wet cloth and dumped her there. I handed her over to my daughter and came back home.

BSMS: What gave you courage to break social norms?

HAMIDA BI: My perspective changed slowly. I felt very awkward for some time. There was no need for me to wear a burkha to show that I am a dignified woman. In our meetings we raised all sorts of women's rights issues like dowry, cheating, and abusing women. A lot of women also changed after they came to the meetings and after they saw their leaders without the burkha.

BSMS: Did you get support from your family?

HAMIDA BI: My family has never opposed my involvement because I have managed to fulfill all my responsibilities. I started my day at five, I cooked for a family of eleven (eight children, two in-laws, and husband) then I headed for the campaign office at ten am and stayed on till eight pm. I came back home and started my stitching work which went on till two am. This was my life because my husband could not earn.

BSMS: What changed your life?

HAMIDA BI: When I joined the campaign my life changed. At that time we collected donations of 25p and 50p in order to run the campaign. We never begged, even when we needed money desperately for court cases. We always got help from our sisters in the center who would even sell their jewelry to get us the money.

BSMS: How do you fight a system?

HAMIDA BI: Everything changes in a woman's life. My daughter was not a bride even for one night. Her in-laws abandoned her. I sued them and fought the case for three years and made them return the dowry that I had already given them. I am a fighter. When life was most chaotic, my husband left me. I could not educate my other children. How was I to send them to school? From where do I get books and uniforms for them? This is the story of my life.

Interview with Rashida Bi, Bhopal, July 2012

A founding member of the Bhopal Gas Peedit Mahila Stationary Karamchari Sangh (Bhopal Gas Affected Women Stationary Workers' Union), Rashida Bi's political commitment is the health care of children with congenital defects, born to gas/contamination affected parents. She runs an organization named Chingari Trust. Dharmesh Shah recorded this oral history with Rashida soon after the July 26, 2012 Bhopal Special Olympics. This event was meant for children with disability caused by gas leak/environmental contamination.

BSMS: Was there a special advantage in doing an Olympics for children with gas-related ailments?

RASHIDA BI: The biggest advantage was the media coverage both national and international. We got more than three hundred hits. We were able to use this opportunity to expose Dow Chemicals, another US-based company, which had merged with Union Carbide in 2001. They refused to take liability for Bhopal. Our effort was to show the innocence of these children and the fact that their lives were destroyed before they came into this world. They want to play cricket, football, like other children. People watching them participating and competing in the games can see their pain.

BSMS: Can you describe the day of the Olympics?

RASHIDA BI: It was a big day. We were able to bring these children to a new place, on this political platform. It gave people goosebumps, and many people cried when they saw that these children were doing this against all odds. They were unable to express their feelings. The audience was spellbound, they wanted to witness every bit of it and be amazed by it. It was a sad day and a happy day. It was also important for us to strengthen our resolve to bring these children to the mainstream and even to the Olympics. I will always remember the day, and this stadium will always be remembered for everything the children achieved.

BSMS: What is your response to the criticism that you have used children for a political agenda?

RASHIDA BI: This was not about the use of children for something political. It was about bringing reality to the forefront. It was not something for which we had to tutor them or something that was forced on them. It was just a different day in their lives. These children experience a life of misery on a daily basis, and we just tried to highlight that through the event. I do not see this as child abuse. It was a display of hope and courage, to show that we were able to do what others could not.

BSMS: What has been the response of the families?

RASHIDA BI: It has given the families a lot of self-confidence. Children who come to Chingari for therapy enjoy themselves. Those who never walked, walked on that day. One child, Aman Quereshi, who never leaves his granny's lap or the bed, tried to crawl that day. His grandmother was amazed to see his behavior, she said this was the first time he was doing this.

BSMS: If the gas tragedy had not happened, where would you have seen yourself?

RASHIDA BI: I would be making bidis, or be at home. Bhopal gas tragedy has taught the women of Bhopal one thing, we have to recognize our own strength. Especially in the Muslim communities, where women are supposed to have children and take care of the family. The Bhopal disaster changed everything because the family itself was destroyed, with the death of the male bread-earners. These thirty years have completely changed my life! Today I can come face-to-face with powerful people and ask them questions.

Interview with Abdul Jabbar Khan, Bhopal, July 16, 2008

A founding member of the Bhopal Gas Affected Women Workers Union (BGPMUS), Jabbar, a lifelong resident of Bhopal, led the resistance movement from the front lines and became a key public face of survivors in India. He was a keen propagator of environmental issues. He fought for clean drinking water and the need to enhance monetary compensation.

BSMS: What got you involved with the fight for justice in Bhopal?

KHAN: I have lived in Bhopal all my life. My father worked in the Bhopal textile mill. He was a freedom fighter. He died in 1986 because of gas exposure. Social work was in my blood. . . . I had experienced extreme poverty. I had faced similar hardships in my education. I have not been to college. After the gas leak I abandoned all my business.

BSMS: Tell us something about your organization?

KHAN: I was the convener of the Women Workers' Union. It was founded by some of the women known to me, who were working in the work sheds set up for economic rehabilitation programs. Twelve months later, an attempt was made to close down the welfare scheme. We succeeded in getting work for more than 2,300 women. Our demand was that all 152 units in the estate should be operational, and only gas victims should be employed there.

BSMS: The Sangathan continues to be active. What inspires you?

KHAN: In the Sangathan I felt we must relate to the everyday hardships faced by the victims. The other survivor groups looked for ways to give issues a political twist. It is important in a movement to have your primary focus on what people want, which in this case was work and compensation. If you win a few victories then you can fight on a bigger political platform. Otherwise you can lose people's trust.

BSMS: How far can you say that every member of your organization shares your belief in political involvement?

KHAN: The middle and the upper middle class live in a world of their own. So they do not empathize with common problems. They take on issues only superficially, or if they are fashionable. The poor will have to fight their own battles because injustice has been done to them. . . . BGPMUS is a mass movement. We have a large number of people turning up for our events. The police keeps a constant vigil, because they can never be certain of what we can do. Between 1986 and 1987 there were attempts to get our organization banned on religious grounds. We were urging women to give up their burkhas. This was seen as un-Islamic.

BSMS: Do you fight for issues other than Bhopal?

KHAN: The movement as a whole has worked very hard towards maintaining communal harmony in Bhopal. It has changed a lot of misconception among communities. We removed the practice of untouchability from the Sangathan. We need a Gandhian model of localized development. We need to make our villages self-sufficient. This is going to be a long struggle. I have spent the most productive part of my life for the movement.

Interview with Satinath Sarangi, or Sathyu, Bhopal

Satinath Sarangi, or Sathyu as he is widely known, is a founder and member of the Bhopal Group for Information and Action and a managing trustee of Sambhavna Trust. A mechanical engineer educated at Banaras Hindu

University, Sathyu came to Bhopal soon after the disaster and has continued to live here. His vision of fighting corporate crime is international in its scope and dimensions. Sathyu took up the cause by raising the all-important question of who cleans up the site of the gas leak. His tag line is "No More Bhopals."[5]

BSMS: What has been your role in the social movement?

SATHYU: From 1985 to 1995, I worked on publications, newsletters, generating information, taking legal actions. More people were suffering chronic exposure-induced illness . . . the failure of the State government was an opportunity to create and legitimate an alternative approach to health care that integrated non-toxic and drug-free therapies . . . and laid stress on community participation.

BSMS: What is the need of the hour after three decades of nonstop activism?

SATHYU: To tell you the truth the battle for justice has just begun! The only way to fight the government and the company is to stop them from spreading misinformation. The solution is to create a "movement friendly" knowledge base that can empower the community. Health care has been the most politically contested issue. Those who are exposed to contaminated ground water must have access to medical guidance, based on factual information.

BSMS: What kind of help are you talking about? Medical science cannot lock horns with activism.

SATHYU: In several countries where Union Carbide has a large presence, environmental rights, human rights, workers and student unions have formed coalitions with the Bhopal movement. I am aware that our limited resources can only provide health care to a fraction of those in need of medical attention. I am also confident that we have influenced those who continue to believe in the dominant system of health care.

Commentary

Some historians have discredited oral history on the grounds that it is simply too personal. However, in a case study of an environmental disaster, the personal is political. This is true of my own personal journey to Bhopal as well. I visited Bhopal for the first time in 2002, exactly eighteen years after the December 3, 1984 disaster. I hired a taxi and requested that the driver take me through the gas- affected colonies.[6] The taxi driver was skeptical. "This place is cursed," he said. "What will you achieve by digging up graves after so many years? Ghosts don't speak!"

I took a break from teaching literature in a classroom and relocated to Bhopal with the idea of doing oral history fieldwork. I had to win people's trust before I could talk to them, which meant distancing myself somewhat from my academic credentials. I began interacting with survivors, listening to their stories, jotting down the conversations, translating these simultaneously from Hindi to English, and transcribing interviews. My interviews were done in spurts beginning in 2003, but, over time, the work embedded me in the social movement that was triggered by the Bhopal tragedy and gave me an insider's view of how this social movement was planned, monitored, and put into action.[7] It gave me the opportunity to become a participant observer in the social/political movement itself, allowing me to multitask as a writer and campaigner for the cause of justice.[8] The idea was to use oral history to empower survivors to tell their own stories in their own words. Given the fact that a vast majority of the Bhopal survivors were illiterate, the communicative power of oral history needed to replace the written script.

The interviews excerpted in this chapter are the product of fieldwork by the Bhopal Survivors' Movement Study. This team organized oral history workshops and identified topics such as the denial of human rights, corporate liability, and the "polluter pays principle" that were to be explored in the interviews. One of the priorities of the interviews was to capture the continued impact of the disaster over the course of the past thirty years as a counter to narratives being spun by the corporation and by the government. This is what made our work subversive. In one of Rashida Bi's interviews she stated, "The disaster was all wrong, but how to manage the disaster was even worse. Our main task was to prevent the government from getting away with false propaganda."

Two texts I use in undergrad-level literature courses are Mary Shelley's *Frankenstein* and Joseph Conrad's *The Heart of Darkness,* classic models of dystopian literature. Much of the literature related to Bhopal read like these frightening novels. They represented a vision of a world that we all seek to avoid. Once a disaster becomes breaking news, it creates a media-generated spinoff of popular human interest stories.[9] Listening to these stories provoked outrage without the resulting commitment to take up the cause for justice. In the case of Bhopal, visual media used disturbing photographs that gripped people's imagination. The narrative of pain became a constant reminder of everything that went wrong from day one.[10]

What made oral history different was that it gave primacy to the art of listening to stories. Empathic listening allowed the interview to enter sensitive areas of the interviewees' experience. It allowed for dialogue and the sharing of high-powered emotions. The more I delved into oral history, the more I discovered its theatrical as well as therapeutic effect.[11]

The social movement in Bhopal used the *padyatra* (walking for a cause) as a form of protest to demand proper rehabilitation for survivors. Many families had lost their only wage-earners, or they had family members who were so ill that most of their earnings went to medical expenses. Many were reduced to complete destitution. The lack of monetary compensation meant they continued to live in the same condition. The middle-class platitude about moving on in life proved to be fake assurance.

The most vulnerable in the aftermath of the disaster were the aged, pregnant women, and newborn babies. Cases of husbands abandoning gas-affected wives were reported. On a much larger scale, prospective grooms rejected young girls with gynecological ailments. In some cases, married women were sent back to their parental homes because they remained too ill to look after the family or had difficulties in conception. The school dropout rates rapidly increased. The narratives described common patterns of behavior—the desire to run away from home, feelings of hopelessness, suicidal tendencies, the inability or lack of desire to work, crying without reason, and violent outbursts. Many of the government officials I interviewed dismissed the complaints as false. It became important for the social movement to penetrate the "corporate veil" and the "closed doors" within the corridors of power. Technocratic and specialized knowledge stood as powerful barriers. These expressive, emotive, homespun oral histories were especially empowering.

Official reporting on disaster management and its impact on government-sponsored welfare schemes were short-sighted and blatantly prejudiced. Media coverage was also fundamentally driven by ratings concerns. My training in literature and the use of figurative language became my strong point. I was able to convey to students who were interested in doing fieldwork, how memories of trauma failed to disturb the urban middle class. They were engaging and listening to narratives that sounded like *someone else's* story. Therefore, the best way to study a social movement was to grasp its political "relevance" as discourse, which is primarily defined as the *language of entitlement*.[12] Surely, the survivors had the right to mourn, protest, and raise slogans against systemic injustice. They were not blaming individuals. For them, the denial of welfare and compensation was the enemy.

The interview method gave the survivors a sense of owning their stories.[13] At the same time, individual renderings of the tragedy had to take into account the universal nature of human calamity. In her 2015 Nobel Prize acceptance speech, Svetlana Alexievich said, "Flaubert called himself a human pen; I would say that I am a human ear. . . . We haven't been able to capture the conversational side of human life. . . . I love the lone human voice."[14] We are all familiar with the definition of oral history as a research methodology, which is both subjective and objective in nature. Each interviewee wants to tell their own stories in their own words. It is this lone voice that needs to be heard by the interviewer. At the same time, the lone voice speaks on behalf of the collective political consciousness of the impacted community. The Movement study decided to give each interviewee a second chance to listen to their recorded voices. They not only listened, but many of them also asked uneasy questions to team members Dharmesh and Tarunima. One such question was whom they blamed more for this tragedy: Was it the government or the company? The majority of women said with unflinching clarity that they blamed the government, because their own nation had betrayed them. In no uncertain terms, it was a denial of rights that they were entitled to.

In this initiative, the Movement study felt the need to create archives with easy access, so that an entire body of literature was made available for those who were interested. In our efforts, we faced many hurdles from institutional research centers. Given the volatile nature of the subject matter, universities and libraries were reluctant to lend space to survivor groups. Both the government and company were determined to protect their turfs against the stalwarts in the social movement to such an extent that the prospect of creating such an archive was virtually ruled out. The Movement study decided to maintain anonymity of the narrators and not share the videotapes in any public forum. Recordings and transcripts were circulated, but the BSMS attempted to make the process fair by granting primary ownership of the data to the survivor groups. These groups had contributed their stories, but also local assistance with language, community events, and understandings of cultural practice.

A chemical disaster such as Bhopal is not merely an "accident." Its legacy is a trail of death, destruction, lack of proper relief measures, and denial of justice. A social movement takes on the onus of pressuring the government to come up with welfare schemes. But in the case history of Bhopal, the accident and its prolonged aftermath were two sides of the same coin. The government made a policy decision to close down many of the welfare schemes. No effort was made to get proper feedback from the people who were using the system. Campaigners used social media platforms such as Twitter and Instagram as methods of communication, with their inherent capacity to go viral. Naturally, none of these devices can replace academic research. At the same time, oral history, as interdisciplinary approach, drew people's attention to what was happening in Bhopal. The long history of "breaking news" on the tragedy delivered through the media had not delivered proper rehabilitation after decades of a continuous battle for justice. Activists have continued to push for restoration. They had distributed posters to increase awareness and held public meetings for the sole purpose of sharing information and planning street actions. The idea was to take stock of marginal voices, without prejudice or rancor. The visibility of the survivors in the public domain became a threat to the rest of Bhopal. It was of local import but also had global repercussions. As Rashidi Bi shared "We know that Dow and other such corporations are responsible for slow and silent Bhopals all over the world. The day we succeed in holding Dow Chemical liable for the continuing disaster in Bhopal it will be good news for ordinary people all over the world."

In 2002, some thirty years after the event, the remains of the Union Carbide factory sit rotting and rusting in the city of Bhopal. *Photograph by Raghu Rai, courtesy of Magnum Photos*

The group leaders and campaigners created a grassroots archive of their own, which included newsletters, photographs, press releases, legal documents, and newspaper clippings. The BSMS used many of these items as "memory triggers" to stimulate the interviewee's memory of events. Each decade became a benchmark for assessing the prolonged impact of the disaster.

Outside environmental activists brought in the sustainable development theory, which created rifts between members of the social movement. One such sensitive issue was the question of getting funds from international sources. An industrial disaster is meant to deal with questions relating to safety, rehabilitation, compensation, scientific expertise, research protocol, and issues of governance. Environmental justice has a cost-benefit side to it that compromised with issues of development and equitable distribution of rights and benefits.

A distinct feature of the struggle for justice in Bhopal was the importance given to primary source materials for purposes of documentation. It became a vital part of people's resistance, which promoted need-based activism. The internet enabled the Bhopal survivor to use mainstream tools of communication to gain visibility. Participating in television talk shows or international forums gave the survivor an empowering sense of sharing the same platform with reputed policymakers. This social domain was innovative and radical in a significant way, in that it asked that the findings of ethnographic research be communicated to the community. Speaking retrieved lost histories from the quagmire of silences and erasure. This process identified the real owners as the "author" of the narratives. It was time for the researcher/ethnographer to retreat from the field, by making way for the real claimants of the stories.

Right from the start, rehabilitation in Bhopal needed technocratic and bureaucratic cover. As a result, both the corporation and the government misused the political immunity granted to them. The best example is the infamous Supreme Court–upheld settlement of claims against the Union Carbide Corporation in December 1989.[15] Environmental contamination became the classic example of how systemic failure works against the best interest of people. Pressing issues of clean-up, medical treatment for the second generation, and bringing the offending corporation to account continued to be contentious matters of debate rather than the all-purpose effort of implementing policy decisions. It was a serious lack of political will on the part of the state that showed up time and time again.

The social movement used the lack of political will to widen its turf. The only way to combat middle-class indifference was to enter the field with an entire ethnographic tradition of sharing and circulating the data. Designated rank and file workers from the survivor organizations carried out a door-to-door campaign and gathered firsthand accounts of how families were coping with adverse conditions. The aim was to challenge misinformation, which denied people's constitutional right to information. Oral history propagated a form of surveillance that kept the community alerted to different government schemes. The survivor groups registered and attended to their complaints and grievances.

The oral history narratives deal with the non-quantifiable nature of loss. Many of the survivors I spoke to said, "We lost everything on the night of third December." Needless to say, there is an element of exaggeration here. The "monstrous night" did not allow for the kind of objectivity that keeps generalizations in check. Oral history has sometimes been discredited on these very grounds. At the same time, orality allows for a dialogue that reenacts trauma in such a way that spoken words, bodily gestures, and emotional outbursts create a site of contestation. Orality gets preserved as recorded speech or written scripts before the authenticity of these narratives is proclaimed.

In the final analysis, what kind of significance can be attached to these narratives, and to what extent do they become innovative action in the promotion of social justice? Each community has its own way of using literature as a memorial which expresses shared loss and acute pain. The lonely, solitary nature of self-expression often gets subsumed by larger social structures, so that the collective emotions of a community are brought under control.

Ethnographic writing has argued that compelling narratives, which can transform a respondent into a subject and not an object of research, create spaces for critical analysis. I believe that we should give credence to the method of "living" through the experience of trauma, without piles of paperwork and the politics of red tape-ism. Only then can we hope to counter the exploitation of people and their voices with relevant research and meaningful activism.

Oral history avoids conclusive remarks. It also creates space for personal narratives. I will end this essay with a personal note that is very close to my heart. My conversations with Rasheeda Bi (we all called her Appa, or elder sister) were prior to the formal interviews the BSMS recorded. We were fighting separate battles, but what was common between us was the shared experience of damaged bodies. I was diagnosed with early onset of

Parkinson's, and her toxin-ridden body seemed to understand the urgency with which I wrote my books, be they academic or fiction. Ours was the collective voice, symbolized by a mutual endorsement of the otherness that artificially demarcated her world from mine.

The more I spoke to Appa, the more I discovered her potential as a team leader. I respected her commitment to the cause; she never questioned my intellectual acumen. We shared the same political orientation. When I asked her the reason for joining the social movement, her answer broke my heart.

> We keep saying remember Bhopal, but there are some things I want to forget. On that night many people came forward to help. The owner of the Straw Factory on Hamidia Road was an ex-army man. He organized a few trucks to take the suffering and dying to hospital. Everybody was pushing and pulling each other to get into the truck. The children were crying in fear. It was then that I noticed that mothers were pushing their favorite sons into the truck, already jam packed with people. They were leaving their daughters behind. It was then that I decided to join the battle for justice. I will fight till my last breath. When mothers are compelled to make such a choice, doomsday has arrived!

Now, when I look back, I realize that ghosts do talk. For them, the past is here and now. The metaphorical language of pain keeps reality at a safe distance. Therefore, the language of trauma goes unheard. Oral history reveals the extent to which reality gets misrepresented at every step. In the process, the true meaning of compensation and rehabilitation gets distorted. As for justice, it becomes a cuss word we all avoid!

Notes

1. From 2007 to 2008, I had the opportunity to participate in an oral history project in Bhopal. Professor Eurig Scandrett, from the Department of Education, Queen Margaret University, Edinburg, UK, was the principal investigator. I joined as an Indian consultant with my own repertoire of oral testimonials. This formed the Bhopal Survivors' Movement Study, which then hired two research assistants based in Bhopal: Dharmesh Shah and Tarunima Sen. For more information on this effort and their project design, see Bhopal Survivors' Movement Study, *Bhopal Survivors Speak: Emergent Voices from a People's Movement* (Edinburg: Word Power Books, 2009).

2. Justice Krishna Iyer coined the word *Bhoposhima*. It was not just a figurative use of language. He was talking about the "ongoing aftermath" of an industrial disaster. See "Bhoposhima: Crime without Punishment. Case for Crisis Management," *Economic and Political Weekly* 26, no. 47 (November 23, 1991): 2705–13.
3. For a detailed analysis of how corporations like Carbide/Dow get away with mass scale destruction of human life, see Marshal B. Clinard and Peter C. Yeager, *Corporate Crime* (New Brunswick, HJ: Transaction, 2006). Another excellent analysis is David Dembo, Ward Morehouse, and Lucinda Wykle, *Abuse of Power: Social Performance of Multinational Corporations: The Case of Union Carbide* (New York: New Horizons Press, 1990).
4. The Movement study carried out the interviews with the idea of people's involvement with data collection. As an oral history case study, the idea was to allow the interviewees to express their opinions with minimal intrusion on the part of the interviewer/researcher.
5. Although interviewed by the Bhopal Survivors' Movement Study Group, this exchange came from a follow-up email from Sathyu in September 2008.
6. It is common in India to refer to place or locality where a group of people with similar interests live together as a "colony."
7. I received a prestigious three-year fellowship (2005–2008) from the Nehru Memorial Museum and Library (NMML) Teen Murti, New Delhi. It proved to be highly beneficial for me. I could use my designation to get appointments with government officials, doctors, and politicians.
8. The Bhopal Survivors' Movement Study conducted forty interviews using high-end recording devices for audio and video taping of the interviews. They hired professional translators for converting the Hindi scripts into English. The tapes were made available in CD form with limited circulation. A copy of the Hindi CD was given to each interviewee. The hard copy in scripted form was meant for wider circulation.
9. Notable were the kind of articles published in *India Today* and popular magazines like *Illustrated Weekly*. This form of storytelling combined investigative journalism and popular writing. It created the polemical space for gathering oral testimonials, but not under the rubric of doing serious research and therefore not taken seriously by academia.
10. Several studies have been done on the kind of stereotyping that goes into defining trauma victims in relation to issues of class conflict, poverty, urbanization, and environmental politics. See Vandana Shiva, *Staying Alive: Women, Ecology and Survival in India* (New Delhi: Kali for Women, 1988) and Meena Menon, "Bhopal: Poisoned Lives," in *Bhopal Revisited, The Hindu Survey of the Environment* (2003). But the most comprehensive study on Bhopal is Tara Jones, *Corporate Killing: Bhopals will Happen* (London: Free Association Books, 1988).
11. An interesting feature of visual modes of representation is the study of photography in relation to capturing the many "faces" of human suffering in the context of post-industrial disaster. See photographs on the Bhopal Disaster by Raghu Rai, Pablo Bartholomew, and Maud Dorr. They are available on the Bhopal websites www.bhopal.net.

12. A full-length ethnographic study which I found useful, though I disagreed with many of its basic premises, was Kim Fortun, *Advocacy after Bhopal: Environmentalism, Disasters and New Global Orders* (Chicago: Chicago University Press, 2001).
13. See Rama Lakshmi, "Curating a Bhopal People's Movement: An Opportunity for Indian Museums" *Curator: The Museum Journal* 55, no. 1, Special Issue: Communities and Museums (January 2012): 35–50.
14. Sventlana Alexievich, "The Nobel Prize in Literature 2015," https://www.nobelprize.org/prizes/literature/2015/alexievich/lecture/.
15. The settlement amount was fixed at 470 million dollars as final compensation that UCC agreed to pay. Three decades later, the cost of environmental degradation and the necessary remedial measures was set aside by using the settlement as a bargaining tool. No wonder Bhopal remains the biggest travesty of justice in modern India.

5

Floating Reed Islands

Gendered Stories of Resilience during Ecological Disaster in the Mara Region, Tanzania

Jan Bender Shetler

Oral histories by Jan Bender Shetler with elderly women of the Mara Region of Tanzania, August–November 2010

The interviews transcribed here were gathered from elderly women about the time of famine in the late nineteenth century, although many of the accounts do not mention that context specifically. They provide gendered insights into how people coped with hunger and environmental collapse. The stories are related to the narrator's grandmother or great-grandmother. Most of them are from an area of the Mara Region on the southeastern shore of Lake Victoria, just south of the Kenya border. The Mara Region is made up of some fifteen (or sixty, depending on how you count) different small ethnic groups that never coalesced into larger pan-ethnic entities, including groups from three major language families (East Nyanza Bantu, Southern Nilotic Tatoga, and Western Nilotic Luo) A few of the interviews come from my earlier research in 1995 from the region's interior. Taken as a whole, these gendered oral histories demonstrate that women played a central role in responding to famine conditions by making and keeping alive regional networks of survival through their memory.

The first stories rarely talk about the context of famine but rather have the odd similarity of being stories about women accidently floating out onto Lake Victoria on an island of reeds until they came to a new home in the Mara Region. Women from the Ruri ethnic group in the South Lake talk about their grandmothers who arrived there floating on reed islands that had broken off across Lake Victoria. A deeper analysis of these stories demonstrates that they took place during a time of famine in the late nineteenth century, and the women married into the communities where they landed.

Jan Bender Shetler, *Floating Reed Islands* In: *Oral History and the Environment.* Edited by Stephen M. Sloan and Mark Cave, Oxford University Press. © Oxford University Press 2022. DOI: 10.1093/oso/9780190684969.003.0006

Interview with Nyasanje Marita Masau, Nyamweko Nyashenene Maria Masau, and Kasigwa Nyafuru Sarai Majangu, Ruri, September 10, 2010, Bwai

KASIGWA: The Wehunjo clan on my mother's side came from Uzinza, south across the lake. Two sisters were gathering grasses to weave baskets along the lake, and when they pulled hard on the reeds an island of matted reeds broke off from the shore and floated out into the lake. They stayed on the water so long that their family had a funeral for them at home, thinking they were eaten by a crocodile. They traveled for two weeks and had no hair left on their heads because they had no other food and ate it for sustenance. They landed in Kwaya [Mara Region] and were taken in by a person who gave them something to eat and helped them to regain their strength. The younger sister, Nyasinde, stayed in Kwaya, while the older sister Umbere went to Ruri [Mara], where she married Manyama and had three daughters. One of those girls, Nyanjegere, gave birth to my grandmother who gave birth to my mother. Umbere's children married to Zanaki, Butiama, and Ukerewe [all in the Mara Region]. We go to these places to greet our relatives and they come to see us.

Interview with Nyasega Rosalia Masaju, Ruri, September 10, 2010, Bwai

NYASEGA: My great-grandmother on my father's side was from the Mara River. Nyasega was at her father's house. Baba Egongoro married Nyasega's mother in Mara. Then there was a problem and Nyasega ran away. Her mother Nyakerenge was left at home with no help. Nyasega went to the river and the mat of reeds broke loose and carried her away, down the river. The reeds left with her on it and floated out into the lake and came to Suguti. The reeds floated here with the daughter, Nyasega.

Interview with Nyasaru Nyasige Naomi Nyamara, Ruri, September 10, 2010, Bwai

NYASARU: The grandmother who gave birth to my father came from Bukoba, across Lake Victoria, floating on an island of reeds, to Ukerewe, or at least his clan was from Ukerewe [a large island off the south shore of the Mara Region]. The Haya people from Bukoba are a "joking clan" (*utani*) with us. My grandfather went to Ukerewe and married her there, her name was Nyanjugu.

Interview with Mtoba Nyamiti Makuna Mang'araria, Ruri, September 11, 2010, Bwai

MTOBA: Makuna was my grandmother, the mother of my father. Nyamiti was his father's sister. She was a girl from Kara Island [off the south coast of the Mara Region] who floated on a reed island to get here. She was bought for food. My father's relatives bought her for food, in the time of my great grandfathers.

Interview with Mkira Kerenge Amina Hasan Sambiji, Ruri, September 11, 2010, Bwai

MKIRA: Kerenge is the name given to me on my mother's side—my great-great-great-grandmother. She came here from Kamasi near Ukerewe, an island. Four women came together from there to Suguti. They traveled here on a floating island of reeds that broke off when they went to collect grass for making baskets. They arrived at Suguti [Mara Region] this way, and people took them into their homes, they were "found" or "picked up" people (*okotwa*) and were not married. They settled here in Bwai among the Shora clan. They came to Matoto's house from Suguti. They lived there for a time and then had news that a brother had come earlier from Ukerewe Island. The sisters came with the reeds to the peninsula. They asked for their brother who was already here. He already had a family here. They stayed with him and got married there. There were others who came here from other places. They brought Luo and Ikizu from Shirati [Mara Region]. They were bought with sorghum. They went in boats with sorghum to Luo and Ikizu. This was during the time of my grandmother on my mother's side. It was during a time of hunger when a half *debe* of food would buy a child. They wanted food and would sell their children. The Ruri wanted the children to marry and to adopt into the family. The Luo were known as "Wagaya," a bought person.

The next set of stories is more explicitly in the time of famine from the perspective of those communities that took in famine victims. In the Mara South Lake region, members of the Ruri ethnic group recount stories of ancestors stolen from Luo-speaking areas of the Mara North Lake. They describe men who lured the children into boats with food during a famine. Luo and Ruri languages are from completely distinct and not mutually understandable language families. Other stories about the taking of children in boats talk about this as pawning children for food during a famine. Few of these children ever found their way home but were married into the new community. A few accounts talk about their return.

Interview with Nyamambara Juliana Yokobo, Ruri, September 8, 2010, Mugango

NYAMAMBARA: I have Luo blood on both sides of my family even though I am Ruri [South Lake, Mara]. Ruri men went up to Luo areas [North Lake, Mara] with boats and had porridge (*ugali*) on the boats. They invited children to come into the boat to eat and then they would kidnap the children to take them home. In that time people married because of hunger for food. The children went to the lake to fetch water and would be kidnapped. My grandmother came this way as a little girl; the fishermen gave them food and then took them away. My grandfather on my father's side took her as a kidnapped child and raised her until she was big enough to marry.

A LISTENER AT THE INTERVIEW: They would give food for the bride wealth in marriage. At least five children were taken at a time in this way; it happened a lot. Here in Ruri they had plenty of millet and sorghum, while there was a famine in Luo areas.

Interview with Nyasige Wegonjo, Ruri, September 8, 2010, Mugango

NYASIGE: My grandfather, Egonjo, the son of Njeje, was Luo, and I am Ruri. Ruri put him in a boat, a *mtumbwi* canoe, and brought him here. He was still small and it was a time of hunger. His parents got food, and they took the children. They tempted them with food to get into the boat. They came here to be workers. There were not many people here at the time. They mostly went to take children from Luo country and Kuria Kine on the North Lake [Mara].

My grandmother on my mother's side was from Ngoreme [South Interior, Mara]. She was also bought for food in a time of hunger. Here they had a lot of food, while there was hunger there. Sometimes after the famine the parents would try to come and take their children back again. During German rule, porters were forced to carry loads for the Chief to the next stop. When the porters from Ngoreme were resting at the Ruri Chief's house before returning, they happened to see a woman bringing water and recognized her as their sister Nyabise, who had been taken from them. They went home and told their family that she was with the Kara [Ruri]. So people came from her home to see and claim her, even though she was already married there. They called for her to come out until her husband welcomed them inside and made a feast, butchering a cow to welcome her relatives. But when her husband found out that they wanted to take Nyabise back he asked whether they wanted him to pay the bride wealth for her. At first they refused and said they just wanted to take her home, but later agreed to take the bride wealth and went home to Maji Moto in Ngoreme. From then on they visited back and forth with her, and she went and stayed with them for a year. But she remained in Ruri as the wife of my grandfather.

Interview with Nyafuru Zura Marigeri, Ruri, September 9, 2010, Mugango

NYAFURU: I was from a Ruri lineage that took slaves. The father of my grandfather was from Ngoreme. My mother was born in Ngoreme and taken in time of hunger. They went to Ngoreme and bought the children for food. They were adopted as children into families here. My grandfather ran away back to Ngoreme, after he was here a long time and had two children and a wife. They came after him and said he should not leave the children, so he came back here to be with his wife and children. Ngoreme had too much sun and the millet died. It got better for them later, during the time of the Germans.

Interview with Chausiku Marita Kwambara, Ruri, September 9, 2010, Kuruwaki

CHAUSIKU: My grandmother who was my mother's mother was from Simbiti, in Suba along the North Lake [Mara]. The Ruri came with food and fooled the children and took them away in boats; many children came this way. It happened a long time ago, I don't know when, maybe the time of the Germans. They were taken in *mitumbwi* canoes. The Ruri wanted to get more people here. So they bought people who would not have any kin. Because they were without kinship they were like slaves, dependents. Whoever bought them had authority to marry them to whoever they wanted. They became like children of the house and would be taken care of very well. My grandmother was married here in this way.

The next set of stories is from the perspective of Luo-speaking North Lake people whose children were taken during times famine, most of them around 1890.

Their stories describe the Great Hunger of Eating Livestock Hides. These accounts relate how severe the hunger was and how people coped. The desperate nature of their circumstances is obvious in these stories.

Interview with Priscilla Philemon Ageke, Luo-Kenya-Kamageta in Luo-Kakseru, September 14, 2010, Shirati Kabwana

PRISCILLA: My grandmother said there was a big hunger when all the food was finished. They burned livestock hides and boiled them in water to get some nutrition. My grandmother said it was in the time of her mother, so my great-grandmother. They did not go anywhere to get help because there was nowhere to go. If you got food, you would be killed by those who came to take your food. Those who had food came to buy children or cattle if they had any. If you are starving you do what you need to do. They took both girls and boys and they never came back. When they were sold we did not know where they went. I don't know what people came or who they were. It was just commerce, it could be anyone. I don't know if it was only Luo who had a famine then.

During the time of hunger some walked in with food to trade. People from Ukerewe came to take children. Someone said that children were taken at that time. But I only heard that they were bought with food from the boats. The person who said this knew where children were taken to in Ukerewe. One of these came back to visit later in Kamageta [North Lake, Mara]. He was up to buy fish and came home to visit his mother. He was already old before I was born, but because of this connection people kept going there to marry long after this. One young girl married there in Ukerewe, but after her husband died she moved back here. The big hunger when these children left was called the Hunger of Eating Livestock Hides; there were other smaller, more common ones, but this one was the worst.

Interview with Tereza Aneta, Luo-Ugu in Luo-Kakseru, September 15, 2010, Shirati Nyakina

TEREZA: Both the mother of my grandmother and my grandmother were in the bad famine. They farmed to help themselves. The Kara [South Lake, Mara] came here in boats to take children for food. They brought food with them and took the children; the parents got the food. No one stole them. There were none in our family that went that way, it was so long ago.

Interview with Amina Pangani, Luo-Kenya, September 15, 2010, Shirati Nyakina

AMINA: I was born during a time called the Hunger of Ngweso (Locust). The government brought food then. But an early hunger was called the Hunger of Eating Livestock Hides. They didn't go anywhere to get food, everyone just died. But some came to take children; they brought food and took them. Even the Kara came in boats; they wanted girls. From my family some were taken. My uncle, Odongo, was in a war with others at the time of hunger.

Another set of stories that women tell in many sites in the Mara Region narrate ancestors who left to find a new home during a time of famine, upon becoming pregnant before marriage, or undergoing circumcision. They are referred to in these stories as women "picked up" or found by the roadside.

Interview with Kasigwa Nyafuru Sarai Majangu, Ruri, September 10, 2010, Bwai

KASIGWA: One of my names is from my great-grandmother Nyafuru Rugemba. She was from the Island of Kara [off the south shore of the Mara Region] and got pregnant while she was still at home as a girl. The Kara did not think a girl should give birth before she got married. As a result, the family said that the baby should be left to die in the wilderness. But she ran away and got in a boat to the mainland at Majita [South Lake, Mara]. She came to Buringa, to the clan of the Ringazi, and was taken in by two mothers when she cried that those in the boat said she had been thrown away. They took her to their husband and said, "Look, we found this girl and took her in, we did not pay anything for her but picked her up. We will call her Nyafuru after the fish." The husband called her and asked who she was and why they threw her out. She said, "I am pregnant." Her baby died after she gave birth, but she later married and stayed there. She gave birth to the grandmother who gave birth to my father.

Wakuru Nyakire, Sizaki in Ikizu, September 21, 2010, Nyamuswa

WAKURU: My grandmother who gave birth to my mother was "picked up" (*okotwa*) in a time of hunger. She and her sister left home to escape hunger and her sister died. They were Maasai [South Interior Serengeti, Mara and east] girls and taken in by my grandfather and incorporated into this home. He married her. So they have inherited Maasai culture. She had one child, no more. She said that if she did, they would go back to Maasai country. She was picked up near Guta, still Sizaki [South Lake, Mara] but in the wilderness. They came in a time of famine to find food; otherwise the Maasai came to steal cattle.

Interview with Nyabanane Kirochi, Nata, September 23, 2010, Mbiso

NYABANANE: My father was Kirochi, son of Shoka. I don't know any more about him because he came from Ngoreme [South Interior Mara] with his mother, Nyabitara, when he was a child. She had a child but was not yet married, and left home because of that. It was not good to give birth at home if you were not circumcised, but she was already. My father died when I was just small.

Interview with Esta Mukami Nyangi, Kuria-Kiroba, September 26, 2010, Nyabange

ESTA: My grandfather Buhoro married another wife who was Luo. She ran away herself from home because of hunger and after walking a long way knocked on a man's door one night. When he saw it was a woman he asked who it was. She answered, "It is me, the one you will pick up." So the young man took her into his home. He was not married yet, so the next day he introduced her to his parents. They asked where she had come from. Mugaya [meaning Luo or "slave"], who was already a mature woman, avoided the question by her reply that they were lucky because she knew how to make rain. His parents, seeing that her clothes were worn from her long journey, skinned a goat and made her something new to wear. Then they asked her to make rain and an enormous rain started, making the parents very happy. Her name was Mugaya and the name is now used a lot in the family because she could make rain. Because of the Maasai raids it was too dangerous to return home, and so she stayed and married the man. She came on her own and said, don't ask me where I came from, you have picked me up.

Commentary

On the face of it, these stories told by women about the displacement of their ancestors may be hard to understand as environmental history. The stories are not obviously about famine, but when paired with other oral narratives and historically contextualized during the time of ecological collapse and disaster in the late nineteenth century, they begin to take on new meaning. Their stories are very different from those told by men about this time period that describe how famine and environmental collapse led to the disruption and reformulation of ethnic group identities, rooted in a particular landscape.[1] Men's stories allow us to see how ethnic traditions imagined the landscape as a discrete ethnic homeland. Women, however, who do not even acknowledge that they know "history," are much more concerned about keeping alive the stories of their ancestors that endured hardship coming from distant places so that their communities could prosper through the regional networks they created. Environmental historians seeking to understand community response to drought and ecological change must take a gendered approach to oral tradition. Women's memories of widespread networks provide critical insight into the historical dynamics beyond isolated "tribal"[2] narratives that enabled people to survive and recover from ecological disaster. These narratives evoke images of shifting regional connections, and women's roles as cross-cultural mediators are at the center of strategies for community resilience and security.

These interviews with women took place in 2010, when I returned to the Mara Region where I had been doing interviews on the oral traditions of ethnic groups, mostly with elderly men beginning in 1995. That earlier research resulted in an environmental history of the region based on the core spatial images of men's historical narratives.[3] I went back to investigate women's historical memory because I had been told before that "women did not know history" and only told folk tales and gossip. What I did find from interviewing ninety-five women in five different ethnic groups was that women do preserve and pass on a very different kind of history than men. Male representations of the history as a people rooted in a very particular ethnic place are the bread and butter of the common "tribal histories" of Africa. Because women in this region leave their homes at marriage and go to live with their husbands' family (patrilocal), often in a different language and ethnic group, they have become the memory-keepers of far-flung regional networks critical to the survival of the community during times of famine or conflict. Women thus control a very different set of memories than men, making historical research a gendered activity.

The methodology I used to understand gendered histories had to work around women's denial of knowing any history at all. The category of historical narrative was gendered male, and all agreed that women did not to know history as such. I thus tried different approaches to finding out what other kinds of knowledge about the past women maintained and passed on. I eventually learned that women were most comfortable relating the stories or talking about names and objects passed on to them by their own grandmothers. With much more prompting, I also got them to tell their own life stories. In each community I went with a trusted friend who guided me to women who were respected for their cultural knowledge. I became part of the social networks of my guides, and the women extended their own stories to me from this perspective. I spoke in Swahili, but my guide often translated women's responses from the myriad of Mara languages. Only one category of women stepped outside of the accepted gendered genres of oral tradition to tell public history; that is, women who had taken on an independent male role in the community. This role was accepted throughout the region as an option for women to act as men when their fathers had no son to inherit (*omosimbe*).[4]

Understanding the nature of oral transmission in this region allows the historian to untangle the meaning of these interview excerpts that depict the pervasive stories of displaced ancestors during famine times. According to the scholarship on oral tradition in Africa and elsewhere, oral traditions are built around a set of core images that are elaborated anew as the speaker constructs the story each time it is told. These are different from the memorized stories that are told by professional ballad singers or *griot* in West Africa.[5] In the Mara Region, some elders are better and more respected at telling oral tradition, but everyone participates in hearing and exchanging historical memory. In my earlier work on men's more public oral traditions of ethnic origins, migration, and settlement, I theorized that the core images used in those narratives are spatial in nature, representing particular places and landscapes with social and political meaning for ethnic identity. The core spatial images of the women's stories that appear here, however, are those of forging extensive interethnic networks across the lake and throughout the region by the connections made as a result of the exchange of displaced people in times of famine. Although men also tell famine stories of moving to other places to seek aid, they tell them as a "tribal" narrative of how the ethnic group came to its current home settled by people from other places.

Women, however, tell the particular stories of their grandmothers or grandfathers and remember their birthplaces as critical social knowledge that links them to distant places and people necessary for them to survive and thrive in a new place. In a system of patrilocal marriage patterns, women left their natal homes and joined their husband's community at marriage. They were specifically charged with remembering the networks that those ancestors brought with them because they continued to be important as links to places for help during times of famine. Although they tell very particular stories of individual ancestors, they evoke a larger pattern of regional connections across ethnic groups that were created or activated during the time of disasters. The core spatial images of networks linked by displaced women reaching out beyond the region to adjacent areas provide evidence for the dynamics of change in this era.

Women first acquired this knowledge from their grandmothers in the private and intimate space of the home where they slept from approximately age six to the time they were married. At night, the girls were entertained by their grandmothers' animal fables, teaching them about sex and marriage, or relating the tales of their ancestors. Thus, the oral histories recounted here must be understood as knowledge passed on to girls from their grandmothers in a form that they were ready to hear and accept. Many of the grandmothers' stories instructed the girls on how to survive when they moved to join their future husband in a strange community, perhaps speaking another language. They had to learn the relationships within that new family in order to get help for their daily tasks and negotiate their position in the family, as well as to maintain their relationships back home in case things turned out badly for them. Their very survival depended on learning grandmother's lessons, but the community also valued this knowledge.[6]

Communities throughout East Africa during the late nineteenth century experienced drought, the spread of epidemic disease among humans and livestock, and environmental collapse. This period of ecological disasters resulted from a combination of El Niño climate patterns and the indirect effects of caravan trade (beginning as early as the late eighteenth century and reaching its height in the late nineteenth century) which brought cloth, beads, guns, and wire from the Indian Ocean coast in return for ivory and slaves.[7] The trade also brought smallpox and rinderpest, along with increased competition for land and resources. When communities moved away to survive, tsetse fly bush encroached into areas that had been kept open and resulted in the spread of trypanosomiasis and sleeping sickness.[8] Communities responded to these challenges with a variety of strategies, ranging from gathering wild food to, as a last resort, pawning their children for food or migrating in search of new land. Women were often at the center of these strategies as the ones who went looking for food, the ones who were traded for food, or the ones who were married out as the critical link between communities.[9]

What is most striking about the memories of today's elderly women is the mobility of their pre-colonial grandmothers across long distances, in contrast to their own more limited movement within colonial "tribal" boundaries. The stories of the earlier generation portray a particularly itinerant population of women, with significant effects on the nature of memory that was passed on to their granddaughters. Intermarriage across long distances to other language groups seemed to be almost the norm, or at least preferred, in this context, as families sought to spread out their risks by making connections to distant communities where food was more plentiful. In order to survive, households put a premium on extending their networks into other ecological zones through marriage alliances or as individuals seeking a new home.

Those distant connections often continued after the disasters, with visits for funerals, weddings, and other events. Many of the women I interviewed had memories of visiting the families of their grandmothers in distant

These three related women were interviewed in September 2010 about the trials of their grandmother floating on a reed island into Lake Victoria. *Photograph by author*

places, with decreasing frequency as time went on. Through the women's stories about their grandmothers, it becomes clear that the exchange of women across boundaries was the main strategy for dealing with ecological disasters in the late nineteenth century. As a result of these connections through women, social networks expanded dramatically within the region and beyond, allowing for communities not only to survive but also recover from famine and prosper. Women were the medium of this exchange, but also the ones who remembered this in their stories of ancestors who came in times of hunger.

In their oral histories, women explained how their grandmothers (or grandfathers) came to the Mara Region from a distant place, across the lake or inland in the late nineteenth century. Putting these excerpts together, it becomes clear that all of this movement occurred in a context of famine and that children were exchanged or pawned for food by desperate parents as a last resort. There is a clear message that "They wanted food and would sell their children."[10] Another text says it more bluntly: "the parents got the food; no one stole them [the children]."[11] Yet another woman said, "If you are starving you do what you need to do; they took both girls and boys and they never came back . . . they were sold."[12] In fact, we know from many other sources throughout East Africa that during the late-nineteenth-century droughts, people traded whatever they could for food, including their children, who they assumed would be adopted into new families and find security.[13]

It is significant that it is the North Lake Luo women who mince no words about the need to trade children for food and who bring up the language of "slavery" in relation to children that were taken rather than received. They did not always even know where the people were coming from to get children, identifying them as from Ukerewe, Kara, and Ruri, or generally as South Lake people. They identified the famine when children disappeared as "the Hunger of Eating Livestock Hides" because there was nothing else left.[14] Another Luo woman said that during this hunger, "They didn't go anywhere to get food, everyone just died. But some came to take children . . . they wanted girls."[15] This was also the time of a thriving slave trade around Lake Victoria, which connected in Mwanza, south of the Mara Region, with the caravan routes to the Indian Ocean coast.[16] Drawing on the new concept of slavery from the coast, one woman said that Luo people, the most common victims of trade, "were known as 'Wagaya,' a bought person." A Ruri woman says outright, "I was from a Ruri lineage that took slaves."[17] They use the language of commercial slavery—buying and selling—despite the reality that they were incorporated into new Mara families as children or wives.[18]

Why then do the women who tell these stories about their grandmothers insist on their coming, for example, on floating reed islands rather than being bought for food as we assume they must have been? In fact, embedded within the same stories are often both the tragic narrative of selling children for food as well as that of girls appearing on floating reed islands or running away from home. One of the narrators asserts that her grandmother "floated on an island of reeds to get here," but then goes on to say that "she was bought for food."[19] Another story is about a girl floating down the Mara River on a reed of mats, but the narrator also adds that the girl had a problem at home and ran away.[20] It seems important to the women who tell these stories not to tell them as a simple story of a grandmother traded by her family for food. It is only by pulling apart the stories common to each of the different areas and putting them within the larger regional context of famine that we begin to understand why women told these stories about their grandmothers as a way to maintain the valuable networks produced by these exchanges.

One answer to the question of why women covered up the exchange of food for children might be to shift the cause or blame for displacement to an accident of nature, that is, reeds breaking off from the shore, or to the victim herself, that is, girls who went into the boats to get food or who got pregnant before marriage. Thus, the home community and the parents were not responsible for the loss of girls who disappeared on floating reeds, were kidnapped by men, or ran away from home after getting pregnant. But, oddly enough then, these stories are told by the receiving community, specifically by a grandmother, who would have little motive to absolve the sending community, to her grandchild.

The concern for the receiving community who took in the girls, however, seems to be that they themselves are not depicted as predatory but rather compassionate for welcoming the girls into their homes.[21] The stories often describe how the girls were taken into their new homes using the verb *(ku) okotwa*—to be "found" or to be "picked up," used in the passive form that does not assign responsibility for the action. Many of the stories of girls who were "picked up" say that she came after running away from home when she got pregnant "while she was still at home." In one story the co-wives took her home to their husband, saying, "Look, we found this girl and took her in, we did not pay anything for her but picked her up. We will call her Nyafuru after the fish."[22] It may well be that these girls ran away on their own, or perhaps they went out looking for food, but more likely they were sent by their parents to find food to send home or to relieve the burden of feeding everyone at home. A pregnant girl was more vulnerable to exile when she was not married or circumcised. A similar story states that two Maasai girls left home "to escape hunger" and were "picked up in the wilderness" looking for food.[23] One of these girls running from hunger came to a home and bargained to be taken in because she was a rainmaker and could help them.[24] Women claimed the agency of their grandmothers leaving home on their own rather than being traded away by their parents, and the compassion of the community that "picked them up."

A second reason for these alternate narratives might be that floating reed islands are a device to cut off these children from their natal homes and make them fully part of their new homes. Girls, and sometimes boys, could be adopted or more often married into a family as valuable members because they "had no kin." The first story talks about the girls' family having "a funeral for them at home, thinking they were eaten by a crocodile."[25] Since the girls would not have known about a funeral, one must assume that the story functions to break ties at home with finality. Ruri South Lake communities, who captured children, practiced matrilineal marriage patterns, meaning that a man who began to accumulate wealth could consolidate that wealth only by marrying a woman whose children would directly inherit his wealth, rather than having it claimed by their matrilineal kin.

One Ruri woman said that they needed more people, "So they bought people who would not have any kin . . . because they were without kinship, they were like slaves," or absolutely dependent without kin to defend them. She claimed that the person who bought them had full control over who they married and what they did, although she also stated that they were well taken care of like one of the family.[26] Ruri at this time were becoming wealthy from the caravan trade and connections with Ukerewe Island off the coast of Mara. As one narrator says, "The Ruri wanted the children to marry and to adopt into the family."[27] In a matrilineal society, children follow the lineage of their mother, leaving men without kin also at a disadvantage. One woman narrated the story of her grandfather who was bought as a child and "ran away back to Ngoreme, after he was here a long time and had two children and a wife." He was convinced, however, to come back to his children in Ruri. In this case the "bought" man was treated as a "wife" in a patrilineal society, who could not take her children home if she left the marriage.[28]

This explanation of the floating island stories may help to make sense of the seemingly contradictory fact that the stories of "stealing children" from boats by luring them with food were similarly told by the receiving community. Why would they want to portray their ancestors as kidnappers of children? The stories from the Ruri side do not mention but specifically cover up the seemingly obvious negotiations their ancestors had with Luo parents to exchange children for food. However, the logic of this narrative is that stealing children avoids acknowledging the uncomfortable relationship with the children's parents, which would have included obligations between in-laws who paid bride wealth, or parents who pawned a child, and the recognition of the ongoing rights and responsibilities to their daughter. It is much easier to incorporate a woman without strings attached to her kin if she was "stolen" rather than "married" by a bride wealth of food. A Ruri woman said, "My grandfather on my father's side took her as a kidnapped child and raised her until she was big enough to marry." While someone else added, "They would give food for the bride wealth in marriage. . . . Here in Ruri they had plenty of millet and sorghum, while there was a famine in Luo areas."[29] This supposition is supported by the comment of one woman that, "Sometimes after the famine the parents would try to come and take their children back again."[30]

Although these stories of stealing children, like the floating reed islands, served to dislodge ancestors from their original families, women also subverted this male agenda of wealth consolidation by remembering and keeping connections to those ancestral communities. They may have done this both to preserve the dignity of their grandmothers as honorable women with kin, and also for their own protection. Women who could not return to or keep connection with their natal homes were more vulnerable to domestic abuse without natal kin to support their rights.[31] Mixed into the stories of displaced ancestors there are also a number of stories about disappeared children coming back to visit, or the connections that were forged by relations back to their place of origin. A Luo woman said one disappeared man came to visit when he was there to buy fish and saw his mother. She said that "because of this connection people kept going there [Ukerewe] to marry long after this."[32] One woman said that the other daughters of her great-grandmother "married to Zanaki, Butiama and Ukerewe. . . . We go to these places to greet our relatives and they come to see us."[33] Another story mentions the "joking clan" relationship with Bukoba across the lake as a result of this exchange.[34] One story has the interesting twist that girls who arrive on the floating reeds find "news that a brother had come earlier from Ukerewe" and go to live with him until they marry.[35] The longest story about a reunion with a woman's family describes the negotiations that ensue when her family discovers where she is and wants her to come home. Her husband ends up paying bride wealth to keep her there, and brings her family into a legitimate in-law relationship. After that, the families visited back and forth and she spent a year at home with her natal family.[36] In contrast to men's accounts, women's stories most significantly include these ongoing connections to the origin places of their grandparents.

Men's ethnic stories also talk about ancestors coming from distant places around the lake, but more often tell them as a way to talk about how different clans are related to one another and how they split apart as part of a "tribal narrative." These accounts are most concerned with telling how the ethnic group came to be, and so the famine stories are often submerged in the meta-narrative of the emergence of the ethnic group from diverse origins. The first story in this series has clear echoes of male ethnic stories in its reference to the Wehunjo clan coming from Uzinza across the lake in the person of the two girls on floating reeds, who became the founders of the Ruri and Kwaya ethnic groups.[37] The story about joking clan relations also deals with men's concerns about clan connections.[38] More consistently, women's stories provide an alternate version of famine narratives that emphasize resilience and cross-ethnic connections.

Understanding these stories also depends on an analysis of how men and women within different community structures experienced the effects of famine and ecological disaster in different ways. According to the economist Amartya Sen, famines are not the result of an absolute shortage of food but a failure of "entitlements" for households or particular people within the household. When hunger comes, the most vulnerable suffer more than those with privilege in the society or the household, depending, in part, on the society's gender structures.[39] Historian Megan Vaughan adds to Sen's theory about "entitlements" those that depend on marriage, age, and gender, showing that family relations and kinship patterns are critical in determining who suffers most from famine.[40]

It may be precisely because of the value placed on women for both their production and reproduction that they became the household members valuable enough as wives to be traded for food as a last resort. Sen studied the gendered effects of intra-household entitlement failure by looking at the overall female/male ratio (FMR) in large continental demographics.

Without gender bias, women have a biological advantage and live longer. Europe and the United States have the most favorable FMR ratios for women at 1.05, while China and South Asia have the lowest at .93–.94. Significantly, sub-Saharan Africa has an FMR close to one, demonstrating a lower anti-female bias, which Sen ascribes to African women's greater autonomy in property rights and household control, employment, and mobility. Worldwide, the most disadvantaged women have little access to employment earning, whereas African women are valued for their farm labor. Sub-Saharan Africa has a higher "female participation in 'gainful' activities" than all other developing regions.[41] These statistics, although on a continental scale and more recent than the Mara case, indicate that African societies overall may not have displayed a bias against women or girls in famine entitlements. But even if valued, women in the Mara still became the medium for securing family entitlements to food in distant communities.

To understand the nature of intra-household famine response, the particular gender systems of each society are critical. Many societies favor male children in situations of famine because of the perception that they will contribute more to the household economy.[42] In the case of the Mara Region, marriage patterns determined that a daughter was most valuable for her agricultural work (production) and the bride wealth (reproduction) she would bring the family, but she would eventually leave the family to join her husband's household. There are famine stories of boys dressed as girls to be exchanged for food. A mother likely favored keeping her son at home, since her own welfare and access to land depended on her son's rights in his father's family. Young girls who got pregnant before marriage were also vulnerable since their fathers could not now collect a bride wealth for them. Going off to fend for themselves and find someone to take them in as dependents may have been their best option.[43] Thus, based on kinship systems, in times of famine a daughter provided the entitlement for her father to obtain food for the larger household while at the same time giving the girl a more secure place to live with a wealthy man.

On the demand side, the mid- to late nineteenth-century context in which the caravans from the Indian Ocean coast to Lake Victoria traded in slaves and ivory introduced and even normalized the concept of buying and selling persons as slaves. The entrepôt for this trade in the southeastern lake was the fertile and rain-blessed Ukerewe Island, where Mara people had been going for famine relief in the past. However, with the arrival of the coastal trade in the nineteenth century, Ukerewe began selling slaves to the coast, as well as using more slaves on their own farms to produce food for trade. Ukerewe traders bought slaves from Ruri fishermen, who in turn had brought them from Luo country.[44]

The European slave trade exacerbated and severely distorted a system of "wealth in people" already in place. It is no accident that these stories took place in the context of the introduction of a globalized trade in people and other luxury commodities. Even before the slave trade, wealthy men created a ready demand for children, women, or any strangers, as dependents without kin who sought protection from a patron. By and large, they were incorporated into households for extra labor in the fields or in royal courts as wives, clients, or simply to expand the kin group and the prestige of the patron in a situation of a shortage of labor, not land.[45] In the older patronage system on Ukerewe Island, they waited passively for people to come asking for food and giving up the clan connections of their daughter in return. The coastal slave trade provided an incentive for seeking out places where desperation brought people to exchange their daughters, and for finding new ways to increase production through the labor of these dependents.[46]

Most significantly, these people described in the Mara traditions as "picked up" (whether from floating islands of reeds, enticed into boats, or found in the wilderness) were defined as kinless people who were taken from all that was familiar and detached from their homes and social identity. The conditions of the exchange depended on their giving up their clan connections. Ukerewe patrons preferred Luo slaves whose extremely different Nilotic culture and language, as well as distance from home, made them less likely to be redeemed and more likely to stick out as different.[47] They were incorporated into their new families, first as strangers with marginal status, much more vulnerable to being treated as commodities—but gradually, when they had children, they were incorporated as mothers and kin. In African society the opposite of slavery is not "freedom" but "belonging." However, a wife in Mara who joined the distant community of her husband was also a marginal stranger, although one with valuable natal kin networks, moving her to a somewhat more secure place on the slavery-to-kinship continuum.[48] From the perspective of women, being dislodged from their home might mean vulnerability or the chance to make a new life.

Yet, women refused to forget the kinship ties of their grandmothers and worked to preserve those connections in oral histories passed on to their granddaughters. All of these resilient strategies in times of famine involved women as the connecting link to bring the necessary means of survival to their families and themselves. Because women's stories from the time of disasters concern particular named ancestors rather than clan or ethnic group founders that represent the community, they provide a different picture of late nineteenth and early twentieth-century regional history. What is most important in these stories is not the consolidation of ethnic identity in the context of migration and famine (as in men's stories), but the expansion of social networks connecting people throughout the region and beyond. Not only did ties between ethnic groups within the region become more robust through ties of marriage, but each of the four sub-regions made significant connections to their neighboring regions.[49] In these stories we can now see the dynamics of increased dependency in the form of "bought" or "found" women played out in particular local dynamics. This led to the expansion of regional connections through women's marriages and their memory of past networks. These regional networks literally saved the region during a time of incredible stress, but also allowed for a recovery and even prosperity in the early twentieth century.

In other interviews, I learned more about continuing relationships and intermarriage following the initial exchange of people in a famine. A South Lake woman told a story about her North Lake Luo great-grandmother. She met her South Lake husband when she was part of a Luo party who came to bring a bride to marry in Ruri.[50] A man from the South Interior said that his maternal grandmother came from the North Interior looking for food and got married here, returning to visit her home from time to time.[51] One indication of the extent of the networks formed by famine is a South Lake woman who said that in her family, they had mothers from Ukerewe Island and two other ethnic groups. She said when they visited family in those places, someone might find a bride and take her home, later paying the small bride wealth characteristic of matrilineal societies.[52] A South Mara Kuria woman who married in the South Interior said that she had relatives in other ethnic groups of both North and South Interior from past generations.[53] A North Lake woman said that if a man wanted a bride, he sent a woman to find one for him since women could travel without fear of violence and "were good with words." She even took the bride wealth cattle for him. This woman herself brought six brides from North Lake Gire to her husband's family for marriage and influenced many others.[54]

Women spoke nostalgically about these distant regional connections that they maintained through marriage. In other interviews, a North Lake Luo woman said, "In the past it was good to marry far away because this brought friendship and also a place where you could go to get help." The reason she got married in Shirati was because a woman from there was married in her home in Kadem and convinced her to come here to marry her brother. In summary she noted, "It is good to make friends far away."[55] Another North Lake Luo woman said that you wanted to marry distantly, but close enough to be able to come home to visit often.[56] A Luo woman said that her ancestors married women from many places to make relations with people from those places. She noted:

> They had love between them and would visit and marry one another and help each other. They loved each other. But not anymore, now it is only (*ubinafsi*) self-promotion. Everyone takes care of their own problems; even your own brother won't help you. Money has made it that way. People like money more than other people. No more visiting each other, we don't even know each other.[57]

Even though North Lake Luo were the most enthusiastic about distant marriages, they still said that one had to investigate the family of the suitor through people one knows in the area.[58]

Numerous incidents illustrate that new migrants moving into the area during the time of ecological disaster avoided conflict by marrying to cement alliances and to establish connections to people who were on the land first. A North Lake Kakseru man said that when they came to Kadem, they still spoke a Suba Bantu language, but they wanted to marry Luo girls so they began to take out their lower front teeth to get married and slowly became Luo. Their children born to Luo wives began to speak Luo, and so the language gradually changed.[59] A North Lake Gire woman said that said that she learned Luo from her Luo-speaking mother as the Bantu-speaking community began to assimilate.[60] In the South Lake, the anthropologist Hugo Huber attributed language shifts to the fact that, in the past, many wealthy South Lake Kwaya cattle owners married women from South Interior or bought "slaves" from North Mara.[61] A North Lake Luo-Kakseru man said that they got to be friends with Surwa people who were there first by giving their daughters to each other in marriage.[62] A North Lake Gire (Suba) woman said that once you made a marriage between two communities they could not fight each other anymore because they were now one family.[63] In all of these ways, women played a central role in the survival and prosperity of the region during this period of intense change.

The core spatial images of women's stories from their grandmothers depict widespread cross-ethnic networks both within the region and beyond. While men might abstract these as origin stories from the distant past, their connection to the context of the late-nineteenth-century famine and conflict gives them historical meaning. During the disasters there were practical reasons for women to cross ethnic boundaries in order to increase the possibilities for getting help in times of hunger or to facilitate trade. Their marriages contributed to the formation of a larger regional structure for facilitating cross-ethnic interaction. To solidify this structure, common assumptions developed over time across various language and ethnic groups about marriage, bride wealth, inheritance, and residence after marriage. Gaining access to regional resources like salt, iron, and wilderness products was also facilitated by interethnic marriage ties, much as has been the case in other small-scale societies around the world. As a result, intermarriage served as a significant constraint to regional and "tribal" conflict.[64] In spite of the denial of these connections in the colonial "tribal" account that sought "purity of blood," women's memory of the time of disasters remains significant.

Women's narratives preserve the creative ways that people forged networks to survive and prosper in a time of ecological collapse. For environmental historians, these are critical sources for understanding community resources for resilience, particularly in an era of concern about adaptation to climate change. While some worry that climate change will bring increasing conflict over environmental resources, these narratives suggest strategies of connection to others and mutual support. It is only by paying attention to the ways that oral narratives are gendered that we gain access to different kinds of famine stories that subvert a "tribal" narrative rooted in the agenda of men to consolidate their wealth. Women's stories, however, bring to our attention the stories of their grandmothers who crossed language and cultural boundaries to create networks of exchange and reciprocity across the region that continue today.

Notes

1. Jan Bender Shetler, *Imagining Serengeti: A History of Landscape Memory in Tanzania from Earliest Times to the Present* (Athens: Ohio University Press, 2007), 11–12, 135–8.
2. I put the term "tribal" in quotation marks to point out that this is an invention of the colonial era and not the understanding of ethnic identity that would have been in place in this pre-colonial and early colonial time period.

3. Shetler, *Imagining Serengeti,* 18–25.
4. Jan Bender Shetler, "The Gendered Spaces of Historical Knowledge: Women's Knowledge and Extraordinary Women in the Serengeti District, Tanzania," *International Journal of African Historical Studies* 36, no. 2 (2003): 283–307.
5. Jan Vansina, *Oral Tradition as History* (Madison: University of Wisconsin Press, 1985).
6. Here and elsewhere, the fuller story can be found in Jan Bender Shetler, *Claiming Civic Virtue: Gendered Network Memory in the Mara Region, Tanzania* (Madison: University of Wisconsin Press, 2019).
7. Stephen J. Rockel, *Carriers of Culture: Labor on the Road in Nineteenth-Century East Africa* (Portsmouth, NH: Heinemann, 2006); Gerald W. Hartwig, *The Art of Survival in East Africa: The Kerebe and Long-Distance Trade, 1800–1895* (New York: Africana Publishing, 1976), 66.
8. See Helge Kjekshus, *Ecology Control and Economic Development in East African History: The Case of Tanganyika, 1850–1950* (Portsmouth, NH: Heinemann, 1977); and Shetler, *Imagining Serengeti.*
9. Shetler, *Imagining Serengeti,* 151, 163.
10. Interview by author, Mkira Kerenge Amina Hasan Sambiji, Ruri, September 11, 2010, Bwai.
11. Interview by author, Tereza Aneta, Luo-Ugu in Luo-Kakseru, September 15, 2010, Shirati Nyakina.
12. Interview by author, Priscilla Philemon Ageke, Luo-Kenya-Kamageta in Luo-Kakseru, September 14, 2010, Shirati Kabwana.
13. James Giblin, *The Politics of Environmental Control in Northwestern Tanzania, 1840–1940* (Philadelphia: University of Pennsylvania Press, 1992); Kjekshus, *Ecology Control;* John Iliffe, *A Modern History of Tanganyika* (Cambridge: Cambridge University Press, 1979).
14. Interview by author, Priscilla Philemon Ageke, Luo-Kenya-Kamageta in Luo-Kakseru, September 14, 2010, Shirati Kabwana.
15. Interview by author, Amina Pagani, Luo-Kenya, September 15, 2010, Shirati Nyakina.
16. Hartwig, *The Art of Survival,* 66–71, 80–81, 116; E. A. Chacker, "Early Arab and European Contacts with Ukerewe," *Tanzania Notes and Records* 68 (1968): 75–86.; C. F. Holmes, "Zanzibar Influence at the Southern End of Lake Victoria: The Lake Route," *African Historical Studies* 4, no. 3 (1971): 479–503.
17. Interview by author, Nyafuru Zura Marigeri, Ruri, September 9, 2010, Mugango.
18. Interview by author, Mukami Gugwa, Kuria, September 21, 2010, Nyamuswa.
19. Interview by author, Mtoba Nyamiti Makuna Mang'araria, Ruri, September 11, 2010, Bwai.
20. Interview by author, Nyasega Rosalia Masaju, Ruri, September 10, 2010, Bwai.
21. I use the term "girl" instead of "woman" for these individuals because that is the direct translation of the term used in local languages. A female is called a girl until she is married or circumcised and often until she has her first child. These are understood to be not quite mature women even though some of them come having already given birth to a child. They were traded because they could still be more easily incorporated into a new home as a wife.

22. Interview by author, Nyasanje Marita Masau, Nyamweko Nyashenene Maria Masau and Kasigwa Nyafuru Sarai Majangu, Ruri, September 10, 2010, Bwai.
23. Interview by author, Wakuru Nyakire, Sizaki in Ikizu, September 21, 2010, Nyamuswa.
24. Interview by author, Esta Mukami Nyangi, Kuria-Kiroba, September 26, 2010, Nyabange.
25. Interview by author, Nyasanje Marita Masau, Nyamweko Nyashenene Maria Masau and Kasigwa Nyafuru Sarai Majangu, Ruri, September 10, 2010, Bwai.
26. Interview by author, Chausiku Marita Kwambara, Ruri, September 9, 2010, Kuruwaki.
27. Interview by author, Mkira Kerenge Amina Hasan Sambiji, Ruri, September 11, 2010, Bwai.
28. Interview by author, Nyafuru Zura Marigeri, Ruri, September 9, 2010, Mugango.
29. Interview by author, Nyamambara Juliana Yokobo, Ruri, September 8, 2010, Mugango.
30. Interview by author, Nyasige Wegonjo, Ruri, September 8, 2010, Mugango.
31. See Chapter 3 in Shetler, *Claiming Civic Virtue*, 160–73.
32. Interview author, Priscilla Philemon Ageke, Luo-Kenya-Kamageta in Luo-Kakseru, September 14, 2010, Shirati Kabwana.
33. Interview by author, Nyasanje Marita Masau, Nyamweko Nyashenene Maria Masau and Kasigwa Nyafuru Sarai Majangu, Ruri, September 10, 2010, Bwai.
34. Interview by author, Nyasaru Nyasige Naomi Nyamara, Ruri, September 10, 2010, Bwai.
35. Interview by author, Mkira Kerenge Amina Hasan Sambiji, Ruri, September 11, 2010, Bwai
36. Interview by author, Nyasige Wegonjo, Ruri, September 8, 2010, Mugango.
37. Interview by author, Nyasanje Marita Masau, Nyamweko Nyashenene Maria Masau and Kasigwa Nyafuru Sarai Majangu, Ruri, September 10, 2010, Bwai.
38. Interview by author, Nyasaru Nyasige Naomi Nyamara, Ruri, September 10, 2010, Bwai.
39. Karen Coen Flynn, *Food, Culture, and Survival in an African City* (New York: Palgrave Macmillan, 2005); Amartya Sen, "Gender and Cooperative Conflicts," in *Persistent Inequalities: Women and World Development*, ed. Irene Tinker, 123–49 (Oxford: Oxford University Press, 1990).
40. Megan Vaughan, "Famine Analysis and Family Relations: Nyasaland in 1949," in *The Social Basis of Health and Healing in Africa*, ed. Steven Feierman and John M Janzen, 71–89 (Berkeley: University of California Press 1992), 72, 85.
41. Jean Dreze and Amartya Sen, *Hunger and Public Action* (Oxford: Clarendon Press, 1989), 50–59.
42. Ibid., 55–58.
43. Igor Kopytoff and Suzanne Miers, "African 'Slavery' as an Institution of Marginality," in *Slavery in Africa: Historical and Anthropological Perspectives*, ed. Suzanne Miers and Igor Kopytoff, 3–81 (Madison: University of Wisconsin Press, 1977), 12.
44. Gerald Hartwig, "Changing Forms of Servitude among the Kerebe of Tanzania," in *Slavery in Africa: Historical and Anthropological Perspectives*, ed. Suzanne Miers and Igor Kopytoff (Madison: University of Wisconsin Press, 1977), 261–85.

45. Kopytoff and Miers, "African 'Slavery,'" 12–22.
46. Hartwig, "Changing Forms of Servitude," 282.
47. Ibid., 280.
48. Kopytoff and Miers, "African 'Slavery,'" 12–22.
49. South Lake to Ukerewe and Sukuma, South Interior to Sukuma and Maasai, North Interior to Maasai in what is now Kenya, and North Lake to Luo in what is now South Nyanza, Kenya.
50. Interview by author, Nyamambara Juliana Yokobo, Ruri, September 8, 2010, Mugango.
51. Interview by author, Jackson Manuga Muraga, Zanaki, September 30, 2010, Bumangi.
52. Interview by author, Nyegoro Ludia Magori, Jita in Kuria-Kiroba, September 7, 2010, Nyabange.
53. Interview by author, Mukami Gugwa, Kuria, September 21, 2010, Nyamuswa.
54. Interview by author, Juliana Ariya Saramba, Gire to Luo-Kowak, July 31, 2007, Kowak.
55. Interview by author, Mikal Ng'oya, Luo-Kadem in Luo-Kakseru, September 13, 2010, Shirati Kabwana.
56. Interview by author, Priscilla Philemon Ageke, Luo-Kenya-Kamageta in Luo-Kakseru, September 14, 2010, Shirati Kabwana.
57. Interview by author, Prisca Matera, Luo-Ugu in Luo-Kakseru, September 14, 2010, Shirati Oboke.
58. Interview by author, Turfena Miluo Oyaya, Luo-Kamageta in Luo-Kakseru, September 16, 2010, Shirati Kabwana
59. Interview by author, Hezbon Nguka Mbira, Luo-Kakseru, September 17, 2010, Shirati.
60. Interview by author, Juliana Ariya Saramba, Gire to Luo-Kowak, July 31, 2007, Kowak.
61. Hugo Huber, *Marriage and the Family in Rural Bukwaya (Tanzania)* (Fribourg, Switzerland: University Press, 1973), 14.
62. Interview by author, Jared Okombo Nyambok, Luo-Kakseru, September 17, 2010, Shirati.
63. Interview by author, Juliana Ariya Saramba, Gire to Luo-Kowak, July 31, 2007, Kowak.
64. For example see Aaron Podolfsky, "Contemporary Warfare in the New Guinea Highlands," *Ethnology* 23, no. 2 (1984): 73–87.

6

Fighting through the Fallout

Maternal and Feminist Resistance and the Fukushima Nuclear Disaster

Heidi Hutner

Oral histories by Heidi Hutner with Sachiko Sato of Fukushima City, September 21, 2011, and April 23, 2017

In March 2011, Japan experienced both a magnitude 9.0 earthquake and the meltdown of the Fukushima Daiichi Nuclear Power Plant. This oral history project focuses on the female, maternal, and domestic experience of these events. Women's voices are traditionally silenced in Japanese culture, so these female oral narratives disclose important accounts of the disaster that might not otherwise be recorded. Sachiko Sato, a mother of six and an organic farmer, provides the central point of emphasis. Through her narrative, we gain a deeper understanding of the complex domestic, social, economic, health, and gendered impacts of the of March 11, 2011 accident. Additional Japanese mothers' narratives are included as well; they echo and expand upon the Sato family's experiences.

The initial interview with Sachiko Sato came after her public talk at the Ethical Culture Society in New York City in September 2011, where she appeared with her two youngest children, her seventeen-year-old son Yuki and thirteen-year-old daughter Mina. Sato spoke of her family's experience after The Great Japanese Earthquake on March 11th, the resulting tsunami and Fukushima nuclear disaster, and the ongoing hardships her family faces. She spoke about evacuating her children from their home and her current concern for the safety and fate of Fukushima's children. She described the pain of having to walk away from her family's organic farm, a thirty-year investment that she and her husband had made—a decision that would last "forever" due to the high levels of cesium 137 contamination. Sato and her husband had built their

Heidi Hutner, *Fighting through the Fallout* In: *Oral History and the Environment*. Edited by Stephen M. Sloan and Mark Cave, Oxford University Press. © Oxford University Press 2022. DOI: 10.1093/oso/9780190684969.003.0007 *sustainable farm and educational center from the ground up. But, after that day, it was all taken away.*

Her children spoke of the impact the evacuation had on them. The youngest, Mina, cried about losing her home, friends, and community, and talked about her difficulties living so far from their farming community. Yuki, her son, spoke solemnly about how losing the farm also meant losing his work, present and future; he had helped his mother with everything. In that first meeting with the Satos, the family's grief was palpable, particularly the children's. With fierceness, Sato said she had come to New York as a mother to warn the world about the dangers of radiation and to protest her government's failed attempts to protect the people of Fukushima. During that visit, she also met with various environmental advocacy groups and protested in front of the United Nations, where Prime Minister Naoto Kan spoke about the Fukushima disaster. At the end of the Ethical Culture Society event, I met with Sato and her translator, Aileen Mioko Smith, a mother, grandmother, and executive director of Green Action Japan. Smith is co-author of Minamata: The Story of the Poisoning of a City *and* People Who Chose to Carry the Burden of Courage, *the book that exposed the damaging health effects of mercury pollution in Minamata Bay, Japan.*[1]

"As a mother and farmer who lost my home and land because of the Fukushima nuclear disaster, I had to step up and warn the world." Sachiko Sato joined a demonstration at the UN in New York City September 2011, warning the world about the severity of the Fukushima meltdowns. *Photograph by Linda Pentz Gunter, Beyond Nuclear*

On April 23, 2017, six years after our first meeting, I interviewed Sato for a second time. I was in New York and she was in Fukushima City, where she now lives and works, ninety kilometers from the nuclear disaster site. We spoke on Skype with the assistance of a translator, Rachel Clark.

HUTNER: Sachiko-san, could you describe what the day of March 11, 2011 was like for you and your family?

SATO: March 11 was a Friday. When the earthquake began, I was at work, a place called Helper Station Ohayo, which means something like "Good Morning Helper Station." Helper Station Ohayo, located in Fukushima City, is a facility where we help handicapped and elderly people. At the time of the earthquake, I was at the office doing administrative work. I've never experienced such a long earthquake and it was a big shock. All the staff members gathered to talk and determine what we should do. Then, there was a power outage and our cell phones stopped working. We received information through radios. Some of us went out to visit our clients at home—handicapped people and injured people in the community; we visited these clients individually, one by one, in their households. Some had been evacuated to various public facilities. We visited them where they had power, so we were able to watch television. What we saw on the screens was like scenes from a movie—all those images of the huge tsunami engulfing the houses.

Around 11 p.m. that same day I went home to our family farm. On my way, I filled my car with gas. I don't know why, but the idea that I should fill my gas tank up came to my mind. That was very lucky. I was right. A few days later, we began having heavy gas shortages.

At midnight, I arrived home and I took my youngest son, Yuki, and my youngest daughter, Mina, to my eldest daughter Maya's home in Fukushima City. Maya was married. After the disaster, our home, where my farm was located in Kawabata-machi, had no electricity or running water, and I was worried about the Fukushima nuclear plants and potential accidents. We were not able to get any information about the nuclear power plants, but I kept thinking about the Chernobyl accident and the possibility of a nuclear accident in Fukushima because of the earthquake and tsunami. If the Fukushima nuclear plants were to have a meltdown or to explode, I intended to evacuate my children far from the plants.

On March 12th, we found out that the Fukushima nuclear power plants had exploded. On that same day, I had returned to my farm because we had chickens. I needed to feed them and collect their eggs. I received a call from my older daughter Maya, and she told me what happened to the nuclear power plants. As soon I heard this news, I called one of my friends, Mr. Toshihiro Ito, who lived in Yamagata City, one hundred kilometers away from the nuclear power plant. Luckily, Mr. Ito told me to come to his home, as soon as possible, with my family and friends. I called my daughter Maya and told her to get away immediately and go to my friend's home in Yamagata. Maya told me that her husband could not leave because of his job. I yelled at her, "What is more important, his work or your lives?" She replied, "Just give me one more day." Maya said she needed to wait until her husband negotiated with his bosses to take a few days off from work.

HUTNER: What happened after that first day?

SATO: On the night of the 12th, I didn't go to Maya's home; instead, I spent the night with my friends and colleagues at my second home in Ino-machi, where my house had electrical power. Most places had power outages and everyone was in a state of fear. My friends and colleagues and I had dinner together and discussed what to do. One of the staff members, a young mother, had a baby that had been born in January of that year—still a tiny baby, right? She said that she would like to evacuate with her baby, and I told her to go with the others to Mr. Ito's. One of the other staff members was from Iitate Village, which had been contaminated. She also had two young children. I told her that she should go to Yamagata together with the rest of the group.

On the morning of March 13th, my third boy, second boy, second daughter, and the staff members I just mentioned—all went to Yamagata. My daughter Maya went to Yamagata, also, a day later with her husband and family. The road to Yamagata was very, very congested with the people who were evacuating. Usually, it takes two hours, but it took the group six hours to get there. When they arrived, they called me. When I heard the news, I felt so relieved. Several of my colleagues went to Tokyo, and one family, Miwa and Todoroki (wife and husband), went to Hiroshima, where Miwa's parents live. Miwa is a second-generation Hibakusha.[2]

They had a small baby born just two months prior to the accident. The baby had just been diagnosed with Hashimoto's Disease—possibly this was hereditary as the grandparents were exposed to radiation from the Hiroshima bomb. Ironically, they evacuated from Fukushima to Hiroshima, as they feared being exposed to radiation. There was another staff member from my job who was pregnant at that time. Her name was Hideko Yoshikawa. I also told her that she should evacuate, but she had a grandmother who was terminally ill and about to die, and so Hideko, who also had a toddler, decided to stay with her grandmother—putting herself, her toddler, and her unborn child at risk. I told her to put masks and raincoats on herself and her child whenever she had to go outside.

As for me, I went to Mr. Ito's place in Yamagata on the 17th. I couldn't go right away with the others, because my boss had gone to Tokyo and I couldn't leave our sick and elderly clients—they couldn't evacuate, so someone had to stay. We were very short-staffed. Also, my husband was in India and my children and I needed the income. Whenever I went outside I wore a mask and jacket and hat all the time. I left these coverings outside of whatever buildings I entered; that's the best I could do. My job was very demanding, and I could not leave the people who needed me. I soon decided to rent another friend's house that was located close to Mr. Ito's. My children and some of my staff families from work lived in this new place together. I went there only on the weekends, as it was a two-and-a-half-hour commute from Fukushima City, where I work, to Yamagata. My eldest daughter, Maya, and my second son Kota, who went to Mr. Ito's place, were over there for a week, but then they heard that Kota's and Maya's husband's work resumed. Maya's husband went back to work in Fukushima and Maya stayed in Yamagata; she got her own apartment there. So, the couple began living apart.

HUTNER: Could you describe more about what was happening at your family farm and how this affected where your family would live?

SATO: At the time of the accident, my son Yuki and I had been running our family farm, and my husband was living in India. Yuki left on March 13th with Mina, as I said. On March 31, we sent soil samples from the farm to a laboratory in France to be tested. The test results showed that the radiation levels were high—6,000 becquerels per kilogram. At that time, the government was only doing radiation air tests, not soil tests. I had a friend who was doing other soil testing, so I sent my soil in to be tested, as well. We didn't expect the results to be so high. Because of the high radiation levels, we decided to abandon the farm on April 28th, and the children never went back. No one lives there now. For a long time, I was the only caretaker of the farm, and as we had chickens, I had to go back to take care of them. After that, the Iitate and Yamakiya areas became mandatory evacuation zones. The Onami area in Fukushima City showed the highest level of contamination. My organic farmer friends who used to live near to our farm all left. They never returned. I told you that my farm's soil samples showed the 6,000 becquerels per kilogram, right? At that time the government announced that any farms where they showed over 5,000 becquerels per kilogram should not grow any vegetables or fruit. Even before this announcement was made, I had stopped everything. My husband had come back from India in mid-April and we shut down the whole farm. We killed all 250 chickens.

Our income from farming became zero at that point. My only income source was the job at the Helper Station Ohayo. My work at Ohayo was very, very painstaking because many of the helpers had evacuated. Many of the elderly people and handicapped were not able to evacuate, so we were overloaded with extra work. We also had to dispatch a number of our helpers to the evacuation stations where they needed assistance. Our staff families were broken up—wives living separately from husbands, in Yamagata or Tokyo. I moved my family eventually from Yamagata to Yonezawa City (only an hour from Fukushima City), so that my commute would be shorter and I could be with my youngest children. After closing the farm, my husband went to Okayama; it is far away from Fukushima or Yamagata—to the west of Honshu Island. He purchased another farm there and decided never to come back. I will never go back to our old farm either; I started farming at Takahatamachi and providing healthy food for the patients where I work, for the staff members and their families, and for my family. Let me tell you a little bit more about my family.

During the first four months after March 11, after I had made my two younger children evacuate from our farm, it was very difficult for me to visit them as I had to keep working and my job was far away. Mina and Yuki were living together for four months, and Mina was not in school. They knew how to grow food, as I had taught them these skills on our farm, but they were living alone and not getting along. Once I got the new place in Yonezawa, closer to where I work, I was able to go back home every day and they began to get along. But Mina refused to go back to school. She tried to go to school in Yonezawa, but the day she started she called me from school to say that she would never go back there again. She had fallen behind and they wouldn't accommodate her needs, and they were not doing enough to protect the children from radiation in the food. I decided not to force her to go to school.

During that time, I was doing activities to protect the children in Fukushima from radiation. Mina observed what I was doing and it upset her. "You are protecting the Fukushima children instead of me, your own daughter," Mina told me. "So maybe I should disappear from this world. That would be better because that would eliminate one extra worry for you." Mina had written these words on her public blog, too. A friend of mine read it and said to me: "Don't you think your daughter is more important than the children of Fukushima?"

Despite my concern for Mina, it was impossible for me to stop my activities. When Mina was supposed to begin the last year of middle school, she told me that she wanted to go back to the Kawamata School—the school she had attended before the Fukushima disaster. I was worried about Mina for many reasons and so, although I didn't want her to live in our original home in Kawamata which was now contaminated with radiation, I decided to let her commute from our other apartment in Fukushima City, where my eldest daughter Maya used to live. There, it would be safer in terms of radiation exposure. The radiation levels inside that apartment were 0.05 microsievert[3] per hour, which is not too bad—the concrete walls probably provided a barrier from the higher levels outside. I gave Mina certain conditions: (1) She should never eat lunch or drink the milk provided by the school; (2) If she felt any kind of change in her body, she should let me know. Mina was very happy in her last year of middle school.

Now, my other child, Yuki, remained in Yonezawa in Yamagata prefecture, and he was alone. My workload at the Helper Station Ohayo became more and more intense. I wanted Yuki to come back to help me. At that point, I moved with Yuki to the Fukushima City apartment and we lived with Mina. That's where we live now. After Mina graduated from her middle school, I wanted her to go to a boarding school in Oguni-Machi, Yamagata Prefecture, where there was an independent Christian Academy. It was safer there (less radiation), and we went to visit the school to see if she would like it. On the ride home, Mina said, "I'm sorry, I don't want to go." She began to cry and say that I loved the Fukushima children more than I loved her. I told Mina that she is much more important to me than children in Fukushima and that's why I asked her to evacuate in the night of March 11th in 2011. I told her: "You might not be able to go back to your original home, but it is more important to protect your life; and that's why I asked you to evacuate." In the car, Mina listened to my story with tears in her eyes. After our talk, she decided to take the entrance exam for the boarding school and she got in. Once Mina was enrolled and began living at the school, she told me that she was truly suffering and wondering if the school was the right place for her.

In September 2013, there was a Mass at the Christian Academy. At one point during the service, each student was supposed to tell a story about herself. When it came to Mina's turn, she suddenly understood why she had to attend this school: to be protected from the threat of radiation, and that's what she spoke about. Up until this point, Mina had tried very hard not to think about radiation and the Fukushima disaster, but on that day, she faced the issue. Mina thought about her own life, and her future children's and grandchildren's lives, and she finally understood that the nuclear disaster was the reason she had to go to boarding school. Mina explained all of this to her classmates in her speech. She said that she could not live in denial anymore about radiation. To her classmates and teachers, she said, "You cannot pretend, 'Oh, I don't know anything about nuclear power plants and their dangers.' You're just trying to escape from the reality if you do so. It's a sin to say, 'I don't know.' " Now Mina wants to do whatever she can do to make up for the denial she lived with up until this time. It was a very big moment for her.

I put this story into a little book called, *Letters from Mina*. I will send it to you. I want you to share it with others. Since this experience of telling her radiation story in Mass at school, Mina has been much happier and she has become very, very outgoing. Mina really used to love her life in Kawabata where she grew up. She misses the farm and her friends and living there very much. She plans to find a new place and start a farm life like we used to have. She is nineteen years old now. Yuki, now twenty-three, works with me at the Helper Station Ohayo. Today, the workload at Ohayo keeps mounting and mounting.

HUTNER: What about the other children? What was happening with them during this time?

SATO: There is my son, Masashi. I have not told you about him yet. On the day of March 11, 2011, he was in Tokyo. Since then, he came back to Fukushima. One of his friend's companies in Fukushima was very short-handed, and he came back to help. After this, he got married. He and his wife live in Fukushima City. Maya, who evacuated Fukushima, remained in Yonezawa and got divorced. Her husband had no knowledge, no understanding, no will to learn about the dangers of radiation from the disaster. She remarried a newspaper reporter in her new home. She also had a baby with her new husband. My second son, Kota, was married in 2014. He did not want to raise children in Fukushima after he got married because of the danger of radiation exposures. In 2016, in March, he moved with his wife to Yamagata Prefecture. In March 2017, this young couple had a baby.

HUTNER: What about your husband? Where was he during this time?

SATO: I am divorced from my husband now. Since 2010, my husband had been saying that he wanted to restart a new life somewhere else. Until then, he had been an organic farmer. He wanted to do same thing in a different way and place. He said he didn't want to spend any money. He wanted to form a new community. That's why he was in India at the time of the disaster. I think he would have lived in India at that time if it had been possible. After the earthquake and Fukushima disaster, he moved from India to Okayama. He purchased a piece of land on a mountain called Genkai Shuraku. There, he started a community with other Fukushima evacuees. When my husband first told me that he wanted to start a brand-new community, it sounded to me as if he were abandoning the life that we had built together up until this point. We had been together for thirty years. The Yamanami farm, our home, was regarded as a wonderful place by our neighbors; it was totally sustainable and we produced 90 percent of our food needs. We had many interns and trainees from across the nation who came to live and learn from us. We built it all from scratch: every building, every structure—including our home—and housing for the interns. We knew nothing about building when we started and we had no money. Selling chicken eggs would not pay for a new house or other structures. So, it took us seven years to build our house, which we did by hand, and we basically camped during that time. All the effort that we poured into this farm was completely denied by my husband in 2010, and his desire to leave it all was a tremendous shock to me—much more so than the March 11, 2011 disaster. So, when the people all around me fell apart on March 11, I was the opposite—the worst had already happened to me—and I knew what to do. For the safety of future generations, I believed that we should close down all the nuclear power plants. We have to protect the children from radiation, and that has been my purpose ever since then.

HUTNER: Could you say more about your new purpose of protecting the children from radiation and your concerns about health impacts of radiation in Japan post-disaster?

SATO: In March 2011, after the disaster, I went to my family's cemetery, to see my mother who was sleeping there. My mother passed away twenty years prior to the accident. She actually committed suicide. She had suffered from depression. At the cemetery, I thought about babies and young children and the dangers of radiation to their health. In the year of the Chernobyl accident, both my younger sister and I gave birth. My younger sister's child was born prematurely. This baby was born with water accumulated in the brain, hydrocephalus. At that time, we never thought that this could have been related to the radiation from Chernobyl. In my sister's room at the hospital, two other babies and their mothers shared the space with her. In addition to my sister's baby, there was a baby with a hole in his heart, and there was a super-small preemie baby, only 650 grams, really tiny. Japan was several thousand kilometers away from Chernobyl, so we never thought these birth defects were related to the nuclear accident.

These days, post-Fukushima, I encounter many women and their infants, and many of these babies have heart problems—holes in their hearts. I have met more mothers who had children right after the Chernobyl accident, and they, too, gave birth to babies with heart problems and hydrocephalus. This could be due to the Chernobyl radiation, or to the fathers' exposure to radiation at the nuclear power plants in Japan, or both (many fathers in the area work at the nuclear plants). These concerns are all over Twitter and Facebook; Japanese mothers share information about their children's illnesses and birth defects on social media. I think the officials are trying to cover things up. The few studies that are going on regarding children and thyroid disease, for example, use questionable methods. We have found problems concerning the way the studies are being conducted and their results, and I don't think the numbers are accurate. There have also been strange sudden deaths. Some teenagers, healthy teenagers, adults, and I see this at my work, too—people dropping dead suddenly in the year of the accident. At my job, I see many sick people, very weak people, much more so than before the disaster; and others in our community—many strange sicknesses. The doctors don't know how to diagnose them—or they say they don't; perhaps they are not allowed to say these illnesses are related to radiation. Daycare centers are seeing many more sick children in school; the children born after 2011 are weaker—so my friend tells me who runs a daycare center.

As far as the health of my family, I have it the worst so far. Several times a year I feel a terrible pain starting from my chest through my throat and into the interior of my ear. I have not gone for medical tests even though the doctor says I should. When it comes to the thyroid cancer checkups, I don't let my children go to the one run by the Fukushima Medical University; we built our own clinic called the Fukushima Mutual Clinic. That's the place where I send my children for thyroid checkups.

In our family, we get our food from various sources. The rice comes from Shimane Prefecture and the Takahata area in Yamagata Prefecture. The vegetables, I cultivate locally, because it's impossible to get everything from entirely outside the prefecture. We have no choice. We have to eat some locally produced food. We also have our own radiation test facility that I organized. In our home, yes, we have a Geiger counter.

As my interview with Sato ended, I recalled our first conversation, which took place six years earlier in New York City. She asked me how far I lived from a nuclear power plant. I said I lived about thirty-five miles from Indian Point nuclear power plant, located in Buchanan, New York, just twenty-five miles north of where we were standing. Sato and her translator Smith exchanged knowing glances and spoke to each other in Japanese—their heads bowed as if in serious conference. Then Smith turned to me and said that I should put my own affairs in order—"Are both your job and your home nearby to the plant? If so, you should think about selling your house now; because when there is a nuclear accident you will lose your income and your home and then how will you survive? Have two weeks' worth of water and canned food saved in your basement. Wait two weeks to evacuate, because the roads will be jammed, so it will be safer to stay inside until the roads are all clear. And, then, say goodbye to your home. Forever."[4]

Commentary

On March 11, 2011, the Tohoku region of Japan was struck by an earthquake, measuring nine on the ten-point Richter scale, the most powerful earthquake to hit Japan in recorded history. A treacherous tsunami ensued, resulting in a power outage and the meltdown of three out of the six nuclear reactors at the Fukushima Daiichi Nuclear Power Plant. All three cores melted through the steel of the reactors and made their way into containment vessels. Water contact with the zirconium-clad exterior of the fuel rods caused reactor units 1, 2, 3, and 4 to have hydrogen explosions. As a result of the meltdowns and explosions, massive amounts of radiation escaped into the atmosphere, including cesium 137, tritium, iodine, strontium 90, plutonium, americium, rubidium, and noble gases (argon, xenon 133, and krypton). Twenty percent of the poisonous releases dispersed over mainland Japan, and 80 percent traveled offshore to the Pacific. According to official data, the event contaminated 4,500 square miles with radiation above Japan's (then) allowable exposure limit of one millisievert (mSv) per year.

The Japanese government did not reveal the status of the reactors or the extent to which radiation had been released from the March 11 meltdowns until May 17. During that same period, the government minimized the number of evacuation zones by changing its definition of acceptable radiation from 1 mSv to 20 mSv. Together, these two factors contributed significantly to rising citizen mistrust about radiation safety. Many believe that the elevation of radiation limits put their health at risk, and they remain concerned about the accuracy of information around radiation contamination in the media and elsewhere. The general fear of future nuclear meltdowns and this information-related distrust led to a surge in antinuclear opposition after the Fukushima accident.[5]

While large-scale protests, sit-ins, occupations, and other mass public displays of protest largely subsided within approximately four years after the meltdowns, Japanese citizen mistrust of the government's radiation policy still lingers among those I have interviewed. Mothers continue to file lawsuits over being forced to return to evacuated areas, remain vigilant in food protection for their families, participate in citizen science, take part in health studies and receive medical care at non-government facilities, and persist in their efforts to keep nuclear reactors shut down. For the most part, evacuees have not returned to contaminated prefectures, and the radiation-concerned expats I have interviewed intend to remain overseas. Many families stayed put in questionable areas of Tohoku and Fukushima due to the cost and challenge of moving away. As Aileen Mioko Smith explains, "It's easy for outsiders to say, 'evacuate, move away' or 'don't return home' to those from Fukushima, but finding jobs and new homes may not be possible. Without jobs, how are families to feed their children, and where are they going to live?"[6]

As of March 1, 2017, the Japanese government ordered subsidized evacuees to return to communities with contamination levels of twenty mSv or less, and most government support for such evacuees has ended. Concerned parents continue to demand that the government restore the maximum safety limit to one mSv; they argue that, even if an individual home has been decontaminated, hotspots remain surrounding the house, making it unsafe for children to play outside. Additionally, as Mioko Smith points out, such cleanups are not permanent: there is an endless supply of radiation waste from the disaster, and the wind blows radioactive material back into buildings, homes, and yards.[7]

According to Kendra Ulrich of Greenpeace Japan, women and children have suffered unequal impacts post-disaster; women have experienced increased violence and domestic abuse, divorce, economic hardship, negative health consequences, and social exclusion.[8] Fukushima mothers and their children experience the social stigma of radiation-affiliation—they are the modern-day *Hibakusha*,[9] a term originally used to describe survivors of Hiroshima and Nagasaki. Some women who were pregnant during the bombings in 1945 gave birth to children with congenital defects. As a result, they were seen as unfit for marriage and motherhood.[10] Analyses of radiation health studies by Mary Olson and Arjun Makhijani show that women and children are most vulnerable to radiation exposures, and fetuses are the most vulnerable of all.[11] Adult women are twice as likely to develop cancer from the same exposure to ionizing radiation as adult men, and children are five to seven times as likely to develop cancer from the same exposure to ionizing radiation as adult men (girls more so than boys). Despite our understanding of these statistics, radiation safety standards continued to be measured using the model of a white adult male body—the "Reference Man." So far, no studies account for racial differences in respect to radiation impact.[12] Additionally, the 2006 report by the National Academy of Sciences does not include the long-term effects of internal contamination, and such exposures may have significant negative health consequences.[13] The result and concern for women and children in Japan post-disaster: women and children have a disproportionate health and safety disadvantage in relation to all radiation disaster, home, and workplace exposures.

In *Radiation Brain Moms and Citizen Scientists*, Aya Hirata Kimura looks at the double bind mothers experience in post-Fukushima Japan. If they do not trust the government's commitment to radiation safety (and, in particular, food safety), they are often viewed as hysterical and lacking an education in science. They are dubbed "radiation brain moms."[14] Yet, providing beneficial and safe nutrition and protecting their children's health is significant to the ideology of good mothering in Japan. Cindy Folkers sees this critique of mothers as "radiophobia," a misogynist construct. Radiophobia-baiting dates back to the Cold War era, Folkers suggests, when the US government aimed to persuade the general public that atmospheric bomb testing was safe. Folkers (and Hutner) suggest that labeling mothers and women as radiophobic normalizes ionizing radiation and leads to female self-censoring and silencing. Radiophobia also contributes to the cultural denial of legitimate concerns regarding radiation health safety.[15]

Unexpectedly, the nuclear accident gave rise to a nascent feminism. Large numbers of women responded to the accident by publicly protesting nuclear power, challenging food and medical safety, and, in some cases, evacuating with their children to safer locations against their spouse's will.[16] These separations and differing views between men and women about radiation safety resulted in what is commonly referred to as "atomic divorce."[17] Mioko Smith says, "in general, men don't take radiation dangers as seriously as do the women—men worry more about finances, while women worry about their children's safety—where they play and what they eat. This may be because it's the women who have to deal with the day-to-day of the children's needs."[18] In a highly conformist culture such as Japan, where women are expected to be subservient, silent, and submissive, this female-driven movement against the status quo reflects the deep social impact of the disaster: the nuclear meltdowns led to the radicalization of many women.

A similar radicalization of women took place in the United States in the 1950s, when Dagmar Wilson and Bella Abzug formed the antinuclear group Women Strike for Peace. That group has been credited with stopping atmospheric atomic testing in the United States.[19] Other historical examples include the work of German Green Party feminists, such as Petra Kelly, who linked nuclear weapons with misogyny. After Hiroshima and Nagasaki, and the bombing of Bikini Atoll and the contamination of the Japanese commercial fishing boat, the *Lucky Dragon*, the Hibakusha and mothers (as food-keepers) were radicalized as well. The Hibakusha (both female and male) have served as representative antinuclear and global peacekeepers since the 1950s.[20] In Japan, Mioko Smith and her cohort have been active since the 1980s, breaking gender roles by pushing against traditional male authority in a myriad of strategic ways. After years of antinuclear activism, Mioko Smith says she often refers to the American antinuclear group Women Strike for Peace as a positive model for Japanese activists, "I always tell groups of women: if the mothers of Women Strike For Peace could do it, so can we."[21]

Sachiko Sato is an important example of such female radicalization. She became an antinuclear activist after the Fukushima accident, and this is now her primary focus—to protect children from radiation and to put an end to nuclear power. Sato became outspoken and challenged the dominant ideology of female subservience and trust in male authority—in this case, the government's silencing of citizen radiation concerns and the promoting of nuclear power. In Japanese culture, Sato's many actions are significant, but she is not alone.

Japan is a highly "conformist culture; no one should be allowed to be different," Ikuko Nitta, a single, divorced mother explains. "But, after the disaster, I had to get over my fears and speak out against nuclear power."[22] I interviewed Nitta in 2011, and we corresponded, and spoke again in 2017. As the challenges of the disaster increasingly affected Nitta's family, she became an activist. Nitta and Cathy Iwane (another mother I interviewed in 2011), who lived in Wakayama prefecture after the disaster, did not want their children to eat school lunches, as they were concerned about possible radiation in the food. However, school administrators frowned upon these parents' wishes to control their children's diet, as Japanese children all eat what is served in the school cafeteria. Parents do not break this rule, but Nitta and Iwane did. These two mothers' testimonies are echoed in the stories in *Radiation Brain Mom* and Ian Nash's documentary *A2-BC*, a film exploring the challenges and fears of mothers and children living in the Fukushima area. The mothers in Nash's film struggle with the problems of wanting to protect their children from eating tainted food and attending schools in contaminated areas.[23] Several of the mothers in the film stand up firmly to the school administrators. In all cases, there is the risk of becoming socially outcast for speaking out and doing things differently—both for the mothers and their children. Most of the mothers I have interviewed prioritize protecting their children over fitting in, but that is not true of everyone. Those who are concerned about belonging socially might be less inclined to speak with me, an outsider, and to acknowledge their fears, concerns, or criticism about the government's or nuclear industry's management of the disaster.

After the disaster, Sato and Kaori Izumi created what they called "nature camps," safe places for children to live and spend time away from their homes in contaminated areas. Sato and Izumi called the children's temporary evacuation in safe areas "nature camps" so that parents could avoid being critiqued or judged by family members, friends, and neighbors, as in many communities it was not acceptable to publicly acknowledge one's fear of the radiation. Izumi worked to protect these children while she suffered through terminal cancer, spending her final days as an activist. She invested her own funds in the legal fight to keep the Tomari Nuclear Plant shut down.[24] Izumi, Smith, and hundreds of thousands of mothers, fathers, and grandparents marched repeatedly in Tokyo in 2011 and 2012. Masses camped out in front of the Ministry of Economy, Trade, and Industry for ten months and ten days. Ten months and ten days—the gestation period of a child in the womb, according to the Japanese—was meant to draw attention to the harmful impact of radiation on the fetus in the womb.

Other women I interviewed, including Tomoi Zeimer, Ikuko Nitta, and Cathy Iwane, fought to stop the burning of radioactive rubble across Japan. In 2011, the government announced a plan to burn rubble in various prefectures across the country as a means to dispose of the radioactive waste materials. Mothers delivered an anti-rubble-burning petition to Japanese consulates globally, including locations in Europe and the United States. Mothers argued that this was unsafe and would pollute uncontaminated areas with ionizing radiation. The protest had been devised by Zeimer, a Japanese mother living in New York City, who had adopted a baby from the Fukushima area just before the disaster. Zeimer's immediate and extended family—her parents, two sisters, and their families—lived in Japan; she felt she "had to do something to help" her relatives at home.[25] She was also concerned about her daughter's young biological mother, who attended a contaminated high school after the disaster. Zeimer worked on another antinuclear action as well; with Yuko Tonohora, she led a massive antinuclear street protest in New York City to commemorate the first anniversary of Fukushima on March 11, 2012. In Japan, Iwane, Nitta, and others continued Zeimer's activist work to stop the burning of radioactive rubble. These mothers spoke at local politicians' offices, protested, and wrote letters to officials.

When Iwane and Nitta lost the anti-rubble-burning fight, both decided to leave the country to protect their children. Iwane took her two daughters to the United States, where she held citizenship, while her husband stayed in Japan to work and support his family. In the United States, Iwane continued her activism, joining the efforts to help shut down the failing San Onofre nuclear plant in California. Nitta also decided to leave Japan, although her evacuation route was more circuitous; it began right after March 11, 2011, when she first moved from Fukushima to Kimono Town in the Wakayama

Prefecture. In the family's first new home, Nitta's eldest child Hikaru, a twelve-year-old boy, became exceedingly depressed and anxious. Nitta explains:

> When the one-year memorial for the accident was approaching, the TV broadcast many documentaries about Fukushima, and that stimulated Hikaru's traumatized memories. He was also bullied at school, and one boy called him, "Fukushima boy!"—so he was hurt much at the time. Hikaru often covered himself in a blanket from toe to head and cried in the blanket. He often said, "I can't breathe," when he happened to hear or see news about Fukushima or the earthquake. He also said he wanted to throw raw eggs in a field and scream loudly and crazily. Katsumi and Jun [his younger siblings] were also unhappy about being separated from their friends in Fukushima. They said they were depressed sometimes. But Hikaru had the most difficulty in getting used to the new situation and new friends.[26]

Although Nitta was very careful about the food she provided for her family after the accident, her children still showed signs of potential radiation sickness. It turned out that Nitta had mistakenly given her children contaminated rice brought from her home in Fukushima one month after the accident. She had judged it as safe because the "bag was sealed, all the windows and a fan window of the house were closed during the evacuation, and the rice bag was placed at the corner of my kitchen, which was the farthest place from the kitchen windows.

"But the rice was actually contaminated through the paper bag." After eating the rice, the children had "stomach-aches, diarrhea, ear pain, headaches, runny noses, and rashes." Nitta experienced a "tingling feeling" in her throat, and she was "thirsty all the time," despite drinking a lot of water. Additionally, Nitta's hands became itchy, the skin on her fingers peeled off, and there were many cracks in the skin on her palms. She began to suspect the rice. Nitta had it tested for radiation contamination, and the results confirmed her fears. She "cried in shock" and apologized to her children. Nitta then had two of her children tested for ionizing radiation exposures and the tests showed that both were internally contaminated with cesium 137. When Prime Minister Noda announced he would begin burning radioactive rubble all over Japan, Nitta said, "I felt like I was falling down to a dark deep hole. I did not want my children to be contaminated again. I could not put up with the situation in Japan." She then decided to leave the country altogether. First, she traveled to Malaysia, and then to Winnipeg,

Canada, where she now lives. Nitta says, "Leaving was the best decision I ever made."[27] To date, large groups of Japanese women and their families have evacuated to Malaysia, Canada, and Okinawa, among other locations. There is a growing support network of evacuee mothers and, according to Nitta, they find each other on Twitter and establish group homes and support communities abroad. Nitta has created a group home for evacuees in Winnipeg, Canada, and she was part of such a home in Malaysia as well.[28]

Many mothers and fathers I have interviewed found, and continue to find, solace and support in these Japanese expatriate communities. They help ease the burden of retraining for careers, learning a new language, and assimilating. "Starting over isn't easy," said one father, who now lives in Canada.

> I was a professor in Japan. Here, I'm a cook. At first, it was my wife who really wanted to leave to protect our daughter. We had a good life. I had a very good job. One day, while watching my daughter at a birthday party, it hit me. While some of the children were swimming, my little girl was covered from head to toe in clothing and she wore a facemask. My daughter stood there quietly while the other children were swimming. When I saw my child standing there, unable to play freely, I knew we had to go. What kind of life is that for a child? Soon after, we left Japan.[29]

As the father finished telling me this story, he broke into sobs.

A number of the evacuees I have interviewed do not want to disclose their names for fear of retribution or alienating and harming their families at home. They told me of their children's health issues—nosebleeds and slowed growth. They said their children's health has improved since they left Japan, but parents continue to watch closely for possible cancers and other health issues. As the documentaries *A2-B-C* show, excessive nosebleeds are common in children exposed to radiation, and the children and parents in Nash's film greatly fear their diagnoses of pre-thyroid-cancer disease. All three of Nitta's children are doing well so far, and her eldest son's anxieties and fears about the disaster have largely subsided.

While some parents have evacuated to other countries, Sato continues to stay put in Fukushima City and fight for radiation safety. Since 2011, she has sacrificed her own health to protect and aid children, the disabled, and elderly. Sato continues to grow and provide "safe" food for her community, to participate in citizen science, and to support an independent medical facility. As Sato says, after the Fukushima nuclear accident, protecting the future generations of Japan became her primary mission. As for her youngest daughter Mina, who struggled in the first two years following the disaster, she has anxiety about developing cancer, as cysts have been found on her thyroid gland and she has been diagnosed as "A2," a pre-cancerous condition. Mina plans to build a new Yanamani farm in the future, in a new location. In her letters to her mother from boarding school, Mina writes that it was on the Yanamani farm that she had been "the happiest person on earth," as she followed her mother "with little steps" in the rich "green field[s]." But, Mina says, with the nuclear disaster, "everything was broken." Notably, like her mother and many other women in Japan, Mina has become an antinuclear proponent. In a public talk in 2014, Mina described the Fukushima disaster as "a black rain like Hiroshima." And, she asserted, as "long as there are nuclear power plants in Japan, no Japanese [person] should be ignorant. . . . The black rain may fall upon you or on the beautiful Mount Fuji."[30]

Sachiko Sato's oral narrative and the interviews with the other Japanese mothers inform us of important issues that remain invisible in dominant patriarchal accounts of the March 11, 2011 nuclear disaster. Mainstream records of the accident focus on the physical structure of the Fukushima nuclear plants and the ability or inability to contain the leaking radiation, and on the economics of the nuclear industry and Japan post-Fukushima.

Sato's and other women's nuclear narratives disclose a rising feminist movement and the changing roles of women in Japanese society, as well as the fracturing of the domestic world: divorce, separated family members, lost communities and homes, compromised reproduction and overall health, and contaminated landscapes. We learn about what happens to the young, the aged, the disabled—the traditionally invisible. We uncover tales of diasporic families who have fled Japan and formed new communities bound by international underground female support. We learn about the unequal gender impacts of the disaster, and we see the many ways women act to combat these inequities.

As second-wave American feminists note, the personal is political, meaning what happens to women, children, the elderly, the disabled, the other—what happens in the domestic sphere—matters to all. In Japanese patriarchal culture and in nuclear history, there is all too often a false dichotomy between private and public. Families' needs and experiences must be accounted for in political and corporate nuclear energy realms. Putting these important women's stories into print, making them visible, is a feminist act of resistance. Oral histories give voice to the oppressed.

Notes

1. W. Eugene Smith and Aileen M. Smith, *Minamata: The Story of the Poisoning of a City, and of the People Who Chose to Carry the Burden of Courage* (New York: Henry Holt, 1975).
2. *Hibakusha*: a survivor of the Hiroshima or Nagasaki atomic bombs.
3. A sievert, or SI, is the unit of radiation defined as producing the same biologic effect of high energy X-rays. One sievert is equivalent to one hundred rem. One sievert, if absorbed all at once, will cause severe illness; eight sieverts will cause death. Japan's allowable safety limits have been set to millisieverts (mSv). Twenty mSv and under are now considered officially safe by the Japanese government for long-term exposure (although prior to the nuclear accident, that number had been one mSv). Twenty millisieverts is equivalent to one thousand chest X-rays per year, or three chest X-rays every day per year. Steven Starr, senior scientist with Physicians for Social Responsibility, states that as a result of the twenty millisieverts external exposures, we can expect to see "about 1000 additional cases of cancer in female infants and 500 cases of cancer in infant boys per 100,000 in their age groups. There will be an additional 100 cases of cancer in 30-year-old males per 100,000 in this age peer group." Ingestion of contaminated food and drink poses an even higher risk, and cesium-137, released in significant quantities from the accident, biomagnifies as it moves up the food chain. Predator species (humans), thus will be exposed to high concentrations when eating contaminated milk, fish, and meat. In "The Implication of the Massive Contamination of Japan with Radioactive Cesium" found here https://ratical.org/radiation/Fukushima/StevenStarr.html, Starr also notes that children and fetuses are at greatest health risk. In the 1950s, British epidemiologist Alice Stewart discovered that a single X-ray to the womb doubled the fetus's chance of acquiring cancer as a child; she argued there is no safe dose of radiation. On the life and work of Alice Stewart, see Gayle Greene, *The Woman Who Knew Too Much: Alice Stewart and the Secrets of Radiation* (Ann Arbor: University of Michigan Press, 1999).
4. Sachiko Sato and Aileen Mioko Smith, Interview with Heidi Hutner, New York, NY, September 21, 2011.
5. As of 2022, only five reactors are back on line in Japan: two reactors at Sendai, one reactor at Ikata, and two reactors at Takahama.
6. Aileen Mioko Smith, Interview with Heidi Hutner, Kyoto, February 22, 2014.
7. Smith, Interview, 2014.
8. Kendra Ulrich, *Unequal Impact: Women and Children's Violations and the Fukushima Nuclear Disaster* (Greenpeace Japan, March 2017), https://www.greenpeace.de/sites/www.greenpeace.de/files/publications/20170303_greenpeace_report_fukushima_women_unequal_impact_engl.pdf.
9. Robert Jacobs, "Social Fallout, Marginalization after the Nuclear Meltdown," *Asia Pacific*, November 16, 2013.
10. Becky Alexis-Martin, "The Chernobyl Necklace: The Psychosocial Experiences of Female Radiation Emergency Survivors," *Belgeo* 1 (2015): 1–12.

11. Mary Olson, "Radiation is More Harmful to Women" (https://www.nirs.org/wp-content/uploads/radiation/radhealth/radiationwomen.pdf); Mary Olson, "Gender Matters in the Atomic Age," in *Crisis without End, the Medical and Ecological Consequences of the Fukushima Disaster*, ed. Helen Caldicott (New York: New Press, 2014), 187–92; Arjun Makhijani, "The Standard Reference Man in Radiation Protection Standards and Guidance with Recommendations for Change," December 2008, http://citeseerx.ist.psu.edu/viewdoc/download?doi=10.1.1.394.358&rep=rep1&type=pdf; also, see: Heidi Hutner, "Invisible Victims," in *Ms. Magazine* (Summer 2015): 14–15. On unequal gender and radiation health dangers in animals, see: Philipp Lehmann et al., "Fitness Costs of Increased Cataract Frequency and Cumulative Radiation Dose in Natural Mammalian Populations from Chernobyl," *Scientific Reports* 6: 19974. DOI:10.1038/srep19974, 2016; A. P. Møller et al., "Elevated Mortality among Birds in Chernobyl As Judged from Biased Sex and Age Ratios," *PLoS One* 7(4):e35223, 2012; A. P. Møller et al, "Chernobyl Birds Have Smaller Brains," *Public Library of Science – One* 6 (2): Art. No. e16862. DOI: 10.1371/journal.pone.0016862, 2012; A. P. Møller et al., "Condition, Reproduction and Survival of Barn Swallows from Chernobyl," *Journal of Animal Ecology* 74 (2005): 1102–11. Tim Mousseau explains that the research team is currently studying these effects in Japan post-Fukushima 3/11 and has determined similar findings regarding gender and radiation (greater harm to female animals); in Tim Mousseau, Correspondence with Heidi Hutner, Columbia, SC, June 28, 2015, and April 2, 2017.
12. Makhijani, "Reference Man."
13. Mary Olson, Interview with Heidi Hutner, New York, NY, June 21, 2017.
14. Aya Hirata Kimura, *Radiation Brain Moms and Citizen Scientists: The Gender Politics of Food Contamination after Fukushima* (Durham, NC: Duke University Press, 2016).
15. Cindy Folkers, "Normalizing Radiation Exposure Demeans Women and Kids and Risks Their Health," in *Counterpunch*, March 6, 2017. For more on radiophobia, see Elizabeth J. Bromet, "Lessons Learned from Radiation Disasters," *World Psychiatry* 10 (2011): 83–84; Elizabeth J. Bromet, and J. M. Havenaar, "Psychological and Perceived Health Effects of the Chernobyl Disaster: A 20-year Review," *Health Physics* 93 (2007): 516–21. Also, see Heidi Hutner and Cindy Folkers, "Radiophobia, Nuclear Advocacy, and the Silencing of Women" (forthcoming). Our essay examines the history and origin of the term *radiophobia* and its health, discursive, and gender implications.
16. Heidi Hutner, "Women and Antinuclear Activism after the Fukushima Accident," in *Sustainability Handbook*, ed. Robert Brinkmann (New York: Palgrave, 2018), 283–97.
17. Ulrich, *Unequal Impact*.
18. Smith, Interview, 2014.
19. Ann Swerdlow, *Women Strike for Peace: Traditional Motherhood and Radical Politics in the 1960s* (Chicago: University of Chicago Press, 1993). On my mother Shirley R. Hutner's (and her cohort's) maternalist antinuclear and civil rights work, see Raymond A. Mohl and Matilda Bobbi Graff, *South of the South: Jewish Activists and the Civil Rights Movement in Miami, 1945–1960* (Gainesville: University Press of Florida, 2003).

20. Toshihiro Higuchi, "An Environmental Origin of Antinuclear Activism in Japan, 1954–1963: The Government, the Grassroots Movement, and the Politics of Risk," *Peace & Change* 33 (2008): 333–67.
21. Smith, Interview, 2014.
22. Ikuko Nitta, correspondence with Heidi Hutner, April 23, 2017.
23. *A2-B-C,* Documentary. Director Ian Nash (Tokyo, 2013). Feature Video.
24. Kaori Izumi, Skype interview with Heidi Hutner, Tokyo, November 9, 2011.
25. Tomoi Zeimer, Interview with Heidi Hutner, New York, NY, April 23, 2017.
26. Nitta, Correspondence, 2017.
27. Ibid.
28. Ibid.
29. Anonymous, Skype interview with Heidi Hutner, Anonymous Location, March 23, 2017.
30. Mina Sato, correspondence with Sachiko Sato, May 1, 2014 and October 12, 2014.

7

More than H_2O

Exploring the Biophysical and Social Dimensions of Water

Javier Arce Nazario

Oral history by Javier Arce Nazario with Héctor L. Martínez Rivera, water system manager for the Doña Mayo community, Doña Mayo, Barranquitas, Puerto Rico, February 12, 2015

In 2008 I started my first faculty position, as an assistant professor at the University of Puerto Rico (UPR) at Utuado, in the mountains of the Caribbean island. After a long period working and sometimes living in the Amazon, I was excited about living in the rural regions of my native country and about developing a new research question related to natural resource management in the mountainous parts of the island. The UPR Utuado campus is the smallest public undergraduate institution in Puerto Rico, with fewer than two thousand students. Mostly, the administrators, the faculty, and the facilities' staff all know one another.

One day, as I was getting lunch, a cafeteria worker asked me if I knew how to test water quality, because many of the people in his community were getting sick and thought their tap water might be the cause. He explained that the community obtained water directly from a stream, and there had been recent land development upriver. Water for the majority of Puerto Ricans (97 percent) is supplied by large-scale, government-managed systems, and since I was raised in the city, I had assumed that everyone received this kind of water service. The

This work was supported in part by the National Science Foundation [grant #1151458]. Any opinions, findings, and conclusions or recommendations expressed in this material are those of the author and do not necessarily reflect the views of the granting agency. I would like to acknowledge the help of the many rural community water system members that have shared their knowledge and time, as well as Ayana Arce for editing the translation of the interviews.

Javier Arce Nazario, *More than H_2O* In: *Oral History and the Environment.* Edited by Stephen M. Sloan and Mark Cave, Oxford University Press. DOI: 10.1093/oso/9780190684969.003.0008 *situation seemed like a straightforward case of environmental injustice—why were isolated rural communities in Puerto Rico not being supplied with safe drinking water? However, after talking with a community member from an Utuado neighborhood with a locally managed water system, I began to see that this was not a simple story.*

Since that first interaction with the users and managers of a Puerto Rican community-managed water system in 2009, I have done participant observation of rural community water systems throughout Puerto Rico and recorded interviews with community leaders and members in communities that manage their own water systems. I have also collected social data through household questionnaires, and water quality data through water sampling. In February 2015 I visited Doña Mayo in Barranquitas, Puerto Rico, a community that has long maintained its own, self-managed water system. During that visit, I recorded this interview with community leader Héctor Martínez Rivera, who has lived in Doña Mayo since his birth in 1962.

MARTÍNEZ RIVERA: My name is Héctor Luis Martínez Rivera. I'm from here in this sector. I was born here in a wooden house fifty-three years ago, because I was actually born with a midwife: it was my very grandmother who received me when I was born.

ARCE NAZARIO: And what was the community called?

MARTÍNEZ RIVERA: Community Doña Mayo, in honor of my grandmother. Because she was the one who donated the highest site in the community, which was her land, to make the water tank. She lived to be a hundred and fifteen, and died here. She is one of the people who has lasted the longest in this community. Everything I know and have learned has been through my father, and at the same time through my grandmother, because I met my grandmother personally. I even stayed with her in her home. She would tell me everything about the water system. This community was composed of sixty to seventy families. One looked for the water in wells, that is, in what were called fifty-pound water cans which were the butter cans that used to be fifty pounds, and they prepared those cans. Even I would carry water in those cans. These water cans were prepared, and there was a little freshwater well beyond where my parents' house was, and most of the community came to look for cans of water.

ARCE NAZARIO: So it was a spring?

MARTÍNEZ RIVERA: Yes, and it still exists, but not in the condition it was back then, you know, back then people maintained it well and actually made some little channels formed from the very plantain or banana plant, which they made into a channel and thus, well, each person had his own little channel made of what's called the "wimpy" plantain stem. They cut open a certain part of the banana plant that formed a channel and that channel was put directly into the rock so that the filtered water from that rock came out, and there they filled their cans or their water containers which they brought at those times to supply themselves, because we had no drinking water here. Forty-five to fifty years ago, perhaps more, the government performed a study to see if they could bring drinking water to this sector. Because of our altitude, the water didn't reach here. Not by gravity, not by pumping, because the pumps they had didn't have enough capacity to reach where we were due to the distance. Well, there was the task of convening a group, including my dad, my grandmother, and another group of people from the community: Don Julio Febus, Don Julio Figueroa, Vidal Concepción, Chito Gonzáles. Those were the pioneers who created an organization as a community. And they set themselves to the task of organizing themselves to find solutions for the water system, to see how they could solve them, because the community was growing and there were already quite a few families, and there was no way to bring drinking water via what was called at that time the Aqueduct Authority.

Martínez Rivera describes the early water infrastructure construction projects undertaken by the residents of Doña Mayo, and the beginning of the community's organization for water management.

MARTÍNEZ RIVERA: Well, that well was the first well we dug. We dug it because it was near green areas and because it was near the river. They were made through some permits from the owner of the estate, which was Mr. Pedro Rivera Malavé, and he donated the site, too, to make the well. Since there were green areas near the river, the well was dug there and when water tests were done they gave about a hundred gallons per minute, that is, it was eighty or ninety gallons, which was quite good in those times. They connected the pumping system, starting with a three-horsepower single-phase pump which flowed into the tank at a rate of about twenty to fifteen gallons or so and filled the tank, for the amount of people that were there at that time. It filled the tank and had neither a filter system nor a chlorination system. There was nothing back then. The water, as it is taken out of the well, flowed into the tank and from there it was distributed to all of the residents. Already, the residents all had galvanized pipe connections to the water main. All this was achieved through [the United States Department of Agriculture] Rural Development projects. They installed the tubes with my father's yoke of oxen. At the edge of all those roads going up to every residence all the pipe was brought up, and for each resident a connection to the main was made, with a spigot outside the house where a sink went. All the household sinks in all the wooden houses were outside.

Everyone was remodeling his house. Each one was modernizing. Maybe he connected the water to the bathroom, and so the community gradually evolved. To this day, and now the system is about fifty years old, the same system. Except that the system is modernized now. Rural Development, at that time, required the community to be organized; therefore, they enforced a series of regulations, because we had to have a water control. And, actually, the pumping of the water towards the aqueduct entailed an expense of electrical energy. The residents were organized and a meeting was held to pay a water fee. At that time, five dollars were paid as a fee by each resident per month.

Martínez Rivera explains the level of effort required to keep the water system operational and in compliance with federal regulations.

MARTÍNEZ RIVERA: Electricity was going up, that is, the cost of maintaining the system was going up. We met and had to raise the rate to twenty dollars at that time. We went up gradually. I don't charge a salary, nobody in the community charges anything and everything is voluntary, this is for the community's welfare, for everyone living here. If a pipe is broken we all go fix it, if the pump is damaged a group of us will go there to fix the pump. I worked my eight hours sometimes, working there in the police headquarters, and then I would leave and change clothes and put on my work clothes to stay until nine or ten at night fixing pipes, because the pump was located near the river and, constantly, floods would wash away the pipe and it constantly needed repair. All that was being modernized and was gradually renovated, as the years passed, until we raised the level of the pipe in the river so that the river wouldn't carry it away.

. . . when [Hurricane] George in 1998 destroyed the electrical system throughout Puerto Rico, destroyed the electrical system, this community spent three months without power. Three months without light meant there were going to be three months without water—but there weren't. Because, since I worked in the police station, I had a lot of experience and at that time the lieutenants trusted me to direct the FEMA directors to the different sectors that had suffered ruptures or significant damages, and among those inspectors, well, I knew one of them and made contact with one of them, and I told him, I live in a sector where we don't have water. We have the system, but we have no power. And they said to me: "We'll get you a generator until the electrical system is resolved."

Through FEMA, the National Guard came and got me a generator and brought it directly to the well. I got an electrician and we made all the connections, and the community was responsible for supplying the diesel and maintaining the generator. And thus it was done. I connected that generator, I even connected other communities because the generator that FEMA got me was fifty thousand kilowatts and it supplied another community, which had the same system as ours, power from the same generator too. This was because the mayor of Barranquitas met with me to see if I could help them with the water supply. Here the community runs everything, neither the municipality nor the other communities help me. . . .

Well, that whole system, as I have been telling you, has been under my supervision since then, and that is why residents like me have achieved so much for the good of the community. I tried to resign, because it is a burden, you know. The residents held a meeting—but nope. Everyone voted against my resignation. I was trying to leave, and the community wouldn't let me go for the simple reason that the government demands a lot from us, because even the EPA [Environmental Protection Agency] dictates what we have to do. Each aqueduct system has to have a qualified operator which they call aqueduct operators, who are certified as aqueduct operators.

Well, even though I worked, I took my days off, which were sometimes Saturdays and Sundays or weekdays. The days that the Health Department held training through the University of Puerto Rico and other private departments. Yours truly, along with the vice president, had to take that training to qualify as system operators. I went to many, believe me. I went to many training sessions, and even for one of the last training sessions, since I had a work-related responsibility, well, there I was sending one of my children to take one of those trainings. It was quite exhausting, it's voluntary, and we paid for it ourselves. That is, to be certified we had to use that certification as system operators, and almost all of my children here have that certification as drinking water system operators, including myself. Almost none of the other residents did, as they had their commitments and many of them said: "oh look, I can't," and I quote, "look, I can't." Well, I had to comply with the Health Department, so I had to sacrifice myself and I had to sacrifice myself in the sense that I had to drop everything I was doing, because the people I quoted couldn't go, and they called me at the last minute, "look, I can't go, other situations have popped up," and so I had to comply with the Health Department. Because both the Health Department and the EPA now require us to be certified to operate the systems. And I was modernizing every system, as I mentioned before, so that it had a water chlorination system. You know, where we put some tablets weekly, and test the amount of chlorine that the system has. It's all we have. We pay the Health Department, monthly, fifty dollars for them to test the water. Every month, fifty dollars, we pay the Health Department to do the water tests to see if it has coliforms or if it does not have coliforms. That's how we know what water quality we have. Other communities, not being well organized or having coordination, have lost their well.

Martínez Rivera discusses how the community reacted to the Las Bocas project, which was a municipal infrastructure development project in a neighboring town that threatened to impact the Doña Mayo water management system.

MARTÍNEZ RIVERA: By our bad luck we were included in that project without telling us anything. They included the community in the Las Bocas project. The Aqueduct Authority started installing pipes all around the community, installing pipes and looking for sites, until one afternoon I noticed that they put pipes in the direction toward our tank. I went to talk to one of their engineers and asked, "and where are those pipes going?"

"We're going to connect that tank over there to supply the community with water."

"And who do you work for?"

"We subcontract for the Aqueduct Authority, and we are going to take over that aqueduct."

"You're going to take that aqueduct? Ok, no problem." I convened the community directors and said, "We have this situation . . . the Aqueduct Authority wants to take our water system. Our system is good. The

Aqueduct Authority has never served us water and now they want to take our system." So we talked to a lawyer to place an injunction. With some savings that we had, we placed the injunction, which isn't cheap, but we needed to stop the project or otherwise they would have stolen our tank. So we established a lawsuit against the municipality, the construction company, and the water company. It wasn't easy, petitioning for three years in a tribunal. . . .

We don't complain, because we have our own system. And we know how to handle and manage our own water system. For more than fifty years we have done it well, and it has worked for us. Why would we accept something that hasn't worked, or venture into something new, when we know that our system has been working well for more than fifty years, something they haven't been capable of doing? And, look, now they've even built a water tank beside ours. A tank that is unusable, that no one is using anymore. The other day a worker for the Aqueduct Authority told me "Hey, you should fight to get the tank for the community, since we aren't using it." *[Laughter]*

In these remarks, Martínez Rivera expresses his pride in the Doña Mayo water system and its management.

MARTÍNEZ RIVERA: Well, that's where we stand, and even today we continue with the system and through the Health Department we keep doing water tests. And, that's it, I mean, what I'm trying to achieve is that my children, who live here in this community, see that what we have as a water system is not a water system, it's as we all say, a grain of gold. Because besides being water of good quality, it is very good water. Very few communities have managed to fight and sustain what we have achieved to this day as a community. When I saw that people get bills for over a thousand dollars, two thousand dollars, one hundred and fifty-three dollars! All of the residents here, using the water they use, what they pay are twenty dollars.

MARTÍNEZ RIVERA: Another thing the Health Department requires of us are some tests for copper, tests for magnesium—all these tests have to be paid for by the community, all these tests are paid by the community. The tests for copper cost five hundred dollars, and that is paid by the community. And there are other tests that I have to do now, too, because the laboratory that did them for us was in Bayamón and now we have to find a laboratory there in San Juan to test them, because we have to pay there so that they come to take the sample and then we send the tests to the Health Department. But with us they won't have a problem, because our water is high-quality water and here in the fifty years that we have had that well we haven't had an outbreak of a disease or anything carried by water. I mean to say, people get the normal viruses that anyone gets, you have influenza, I don't know, or a gastritis that anybody can catch even from food that you eat outside or at home, but within the water system as such, I would say that our system is one of the best in Puerto Rico.

Here, he discusses the role of the Doña Mayo community water management organization during water shortages.

ARCE NAZARIO: Do you have a plan for times of drought?

MARTÍNEZ RIVERA: Yes, we had a meeting when the big drought came to Puerto Rico. In '94 there was a big drought, so we tasked ourselves with handing out letters, we eliminated what are called little swimming pools, those household pools, we banned the household swimming pools. No cisterns, as I told you, and we were controlling water. Everything like that we controlled, if we controlled the timer we controlled the community water and it was done. The community didn't lack water, because the community never ran out of water, but for people . . . there, we prohibited washing cars, or if they washed them it was once a month, but that thing of washing and doing it every weekend, no, that was forbidden and anyone that was caught was fined. Washing cars that were not owned by community members was prohibited.

Thus, everything that we could figure out that could be harmful for the well-being of the community as far as drinking water was concerned, well, all that was forbidden to the community. There was a control, and thank God we never destroyed the system, the well maintained "gallonage," giving the "gallonage" that we wanted in the community. In fact, other communities came so that we could give them "gallonage" to cover their usage as well as to drink. Because they didn't want to drink the water that the Aqueduct Authority gave them, and we supplied the drinking water for many residents and other communities that asked us for a few gallons of water. We did all of this as a community. Many people came and weren't even charged –it was for free.

ARCE NAZARIO: And on some other occasions as well, besides the drought?

MARTÍNEZ RIVERA: Yes, on some other occasions, too, when there were hurricanes. When we had water but had no electricity, but we had water, as I was telling you before. Well, other people in the community came looking for water and water was given to those people too. To people that we understood were in need, old people, a lot of them, too, we took water directly to their communities, to those where we knew that there were old people who didn't have any. Although they had the system from the Aqueduct Authority, they didn't have the same system we did. We brought water to these people. The fact that we own the system doesn't mean that we are very strict, no, we are also generous with others. When we see the need, we help. We help each other. You never know if we might need it, too.

ARCE NAZARIO: It seems as if this community is well organized. Why do you think that is?

MARTÍNEZ RIVERA: I would tell you that the warmth of the people in the community and the fact that the community appreciates what we have, I think, is what has made us persist for so long. Because there are many professionals. I'm telling you, here there are doctors, and here there are teachers, there are police officers, there are lawyers. That is, all the people who live in the community know what we have, and as I say, it's like a grain of gold. The system we have, we call it a grain of gold, because I believe that the most valuable liquid in the world is water. I believe that without it, no human being can survive, without water. Nobody in the world. And that is why we love this system and appreciate it—because it is water.

Commentary

Héctor Luis Martínez Rivera's interview brings together a multifaceted history of the community values and social changes that have affected a small community water system. Martínez Rivera's interview certainly does not cover all of the possible interactions and experiences that people organized around a community water system might have, but it does exemplify several of the most common aspects of the histories of small community rural water systems that I have worked with. It is also a particularly beautiful interview, beginning and ending with observations that remind us that a water system is not only a simple public utility, but can also be an environmental resource of great value

Two large cement water tanks—one owned by the Puerto Rican central water utilities system, and the other by the local community—stand at the top of a cliff in Doña Mayo in the island's central mountainous region, and community members prepare to clean their tank during a communal work day. The proximity of the two tanks demonstrates that the Doña Mayo community maintains its sovereign water system despite the availability of government water services. *Photograph by author*

for anchoring the identity of the community members. Martínez Rivera's words convey his pride in managing and conserving a natural resource, while also demonstrating a detailed understanding of the management system and of the laws that govern water quality at the local and state level.

The excerpts that I have selected are also especially relevant to many of the academic questions in environmental science. Water is a critical natural resource impacted by climate events, changes in land use, and population growth. Water can be privatized or commercialized, and it has a multifaceted role in public health. Thus, discussions of water services touch on questions of public health, governance, water quality management, and resilience to environmental changes. To highlight these issues, I have focused this commentary on three principal themes: public health, common property, and resilience and adaptability.

Rural community water systems in Puerto Rico have attracted attention as a public health concern since the 1990s, and yet they continue to supply water to 3 percent of the Puerto Rican population.[1] In attempting to characterize the role of these water systems, Martínez Rivera's enigmatic statement that "besides being water of good quality, it is very good water" serves as a useful guide, by reflecting a critical understanding of the concept of water quality and the metrics typically used to assess the impact of water systems on public health and environmental services. The Environmental Protection Agency (EPA), which establishes the environmental policies for US territories as well, includes these community water systems among areas of concern for environmental justice in Puerto Rico.[2] The EPA and public health agencies in Puerto Rico have called attention to community water systems especially by highlighting the problems that many such systems have had complying with the federal Safe Drinking Water Act (SDWA), which defines the treatment, measurement, and reporting requirements of tap water systems. Among the several categories of SDWA compliance failures on the part of rural community water systems, some of the most common are that system managers fail to test the water quality or fail to submit testing reports to the environmental agencies on schedule. In other cases, these water systems do not meet compliance standards because their water treatment methods do not conform to regulations, or because the water inherently has certain characteristics that are unacceptable under SDWA regulations.[3]

The story of compliance failures on the part of the rural community water systems is part of a larger narrative that brings these water systems into the forefront of environmental justice concerns for the EPA and other agencies. Other aspects of this narrative characterize the social composition of communities with self-managed water systems, typically depicting them as poor, isolated, and lacking organizational and physical infrastructure. Together, the narrative paints a devastating picture of communities like Martínez Rivera's as having been unjustly left behind and exposed to grave health risks because they are excluded from the technological progress in water distribution and management.

Martínez Rivera's interview reveals another dimension of the interaction between communities and agencies that is reflected so starkly in EPA and other reports. While he discusses compliance problems and explains that his community's water managers are in constant conversation with the government health agencies, he also communicates a narrative that directly contrasts with the story of injustices and risks. This alternative narrative recounts the community's shared memory that no one has ever become sick from the water, a shared memory that is frequently mentioned in my interviews with communal water system stakeholders from many communities and an obvious root of these communities' pride in their water. Scientists and public health advocates would probably point out the danger in this idea, since it suggests that the community's water is safe and always will be.

But the oral histories I have recorded do not suggest that the community water is a magically safe, ever unchanging resource. Instead, these oral histories (including Martínez Rivera's) show how the community has modified the water management system to counteract possible risks from changes in the natural and constructed landscape. The histories include descriptions of how the communities have organized to move wells or dikes so they could have more water or better water quality. In these descriptions, we find an understanding shared by communities and environmental agencies alike: water consumption and water quality change continually as a result of social and environmental processes.

Data from the Health Department in Puerto Rico corroborate the shared memory mentioned by so many community leaders: currently, not a single record documents any water-related gastrointestinal epidemic that affected users of a rural water system.[4] In contrast, epidemics of gastrointestinal disease that arose in contaminated government water systems have been recorded, which is not necessarily surprising since the government serves many more users than the sum of all community water systems. Gastrointestinal illness is not the only danger that water quality recommendations protect against, however, and there are a few recorded cases of rural community water systems that have become contaminated by chemicals resulting from land-use change, such as agricultural pesticides, fertilizers, or industrial compounds.[5] But again, these risks are not particular to small systems. More extreme health risks have threatened users of government water systems as result of contamination of the source water and of problems in water treatment procedures.[6]

The shared memory of safe drinking water conveyed in oral histories like Martínez Rivera's helps us to see beyond the binary of community- and government-managed water in thinking about water risks. Public health is a major concern for any group dealing with water, and the oral histories reveal dimensions of water management that may be neglected or even undermined by focusing exclusively on water quality variables.[7]

The oral history of Puerto Rican rural community water systems can be read as a history of communities negotiating their local geographies, their economic environment, and political structures to bring safe water to their homes.

It is often a story of agency, conservation, and adapting to change. While agency and social capital are not variables measured by public health officials or environmental agencies, they have long been important foundations of the concept of environmental justice, and Martínez Rivera's oral history shows that agency is an important asset derived from the water system. Even the consumers and managers of systems with serious management problems can simultaneously criticize the water quality and recognize how their systems are important to the identity of the community and to a sense of self-sufficiency. Self-sufficiency and community pride are not necessarily priorities in environmental policy, which is usually designed around compliance to facilitate the application of laws. Measures have been proposed to solve the problem of SDWA noncompliance in small rural water systems in Puerto Rico, specifically, to eliminate these systems and connect the consumer households to large government water systems instead.[8] This action is frequently recommended in EPA reports as a way to bring environmental justice to the affected communities, despite the fact that self-determination is a core value of environmental justice that these actions would paradoxically jeopardize.[9] The description of rural water systems in Puerto Rico found in government agency reports deprecates the value of agency by generalizing their consumers as communities composed of an uneducated or unorganized rural population constantly being exposed to unhealthy water. As a 1997 technical report demonstrates, rural community water systems are characterized as poor communities that barely have organizational resources:

> Puerto Rico has a total population of approximately 3.7 million, 97% of these are served by the government utility, the Puerto Rico Aqueduct and Sewer Authority (PRASA). The remaining 3% (more than 100,000 people) relies on their own rudimentary water systems for drinking water and water of general use. Most of these water systems originated because PRASA lacks the capacity to provide the service to some rural areas due to the extremely difficult access (high altitude and mountainous topography) and the high cost of providing it. In other cases, the nearest PRASA system is working over its design capacity.
>
> These systems, commonly called Non-PRASA systems are subject to compliance with the same laws and regulations that govern the design, operation and maintenance of PRASA systems. Nevertheless their users and operators, mostly rural, isolated and poor communities, barely have the technical, economic and organizational resources necessary to achieve

and maintain compliance. Moreover these communities are continually exposed to serious health risk of contracting water borne diseases.[10]

The narrative represented by this technical report is not altogether wrong: there are systems in very remote places where the community has not become sufficiently organized to avoid serious risks related to the minimal management of water quality. But intervention by the government is not necessarily welcome. Martínez Rivera's story informs us that even his community's very successful water system became the target of a government infrastructure improvement project, and that his community went to great lengths to take legal action to prevent the system from being overtaken and replaced by the Aqueduct Authority.

Oral histories from other community leaders describe communities that once received water from the government system, and as a result of dissatisfaction with the state-managed resource decided to create their own. There are even communities that receive water from two systems: one governmental and one community-managed rural water system. The narrative that portrays a degraded system managed by a disenfranchised social group ignores and obscures the creativity, leadership, and initiative that many community water managers have shown in their efforts to comply with the regulations.

Martínez Rivera reminds us of all the training that he has taken to manage the system and keep it in compliance with regulations. He also reminds us that the community is not the homogenous group of poor, uneducated people represented in the technical reports. He insists that the community is diverse and gains from the diversity of knowledge. In rereading his discussion about the diversity of the community, as well as his descriptions of the technicalities of the water system, I wonder whether terms like *rudimentary, isolated*, and *lacking organizational resources* convey the technocrat's prejudice more than any truth about the Doña Mayo community. Though the concept of the "tragedy of the commons"[11] has been discredited multiple times in environmental research,[12] the idea still lurks behind many of the arguments for the privatization of resources as a more efficient and productive means of providing a service.[13]

At the time of Martínez Rivera's narrative, privatization was not an important part of the political dialogue surrounding water governance in Puerto Rico, but it is a part of a global dialogue about water and property. Communities like Martínez Rivera's have analogues in other countries where water is being privatized,[14] and oral history has a role in the conversation. Oral histories like the one recorded here can show what works and what is difficult in managing water as a community resource, and, especially, what the community gains or values most in controlling their water production.

Martínez Rivera's account presents valuable lessons on governing the commons. The stories of water rationing, sharing with other communities, and responding in teams to emergent situations contradict the dark conclusion of the tragedy of the commons. Instead, the oral history provides us with an understanding of how the limits of a water resource are confronted by the community: when necessary, the community acts to preserve something that everyone values. His description of the community water fees and of water restrictions during droughts provide evidence to support the hypothesis that in a healthy communal system, there should be consistency between the restrictions on access to the resource and the local conditions.[15] The community confronted the increasing demand for water resources caused by population growth by creating management rules, and eventually by deciding to divide the water system into two separately managed community water systems when the size became unmanageable. Martínez Rivera's narrative demonstrates that the community has strategies in place to manage changes in the environment such as droughts and extreme precipitation. Response strategies for these kinds of risk are also important aspects for healthy management of common property.

Recognizing the need to adapt is a key part of the Doña Mayo community's approach to managing the commons. It is also valuable to understand the limitations of community principles and to recognize that leadership and initiative from individuals is an important factor in promoting a proper collaborative environment. Martínez Rivera's interview reminds us how malleable the concepts of community are, as he acknowledges that there have been moments of greater participation and others of less participation. Yet, the interview shows that there is a unifying interest among the community members in conserving the water systems. The descriptions of the many times he had to take responsibility for required training while other community members found last-minute excuses to not participate do not reflect the utopian view of community enterprise in which everyone participates generously at all times, but they do bring us closer to acknowledging "real" community dynamics. Social psychologist Ester Wiesenfeld reminds us that "variations among people in the degree and form of involvement, activity, participation, and commitment to the community should be expected. The *community concept* of *community* needs to protect and value as positive the differences among its members."[16]

Despite these realities, the image of unified communities working together is commonly prescribed as necessary for producing positive environmental outcomes, a false image that may disillusion those that would like to promote communal environmental efforts. Martínez Rivera's oral history is not unlike that of many environmentally focused interviews I have recorded. These stories often remind us that communities, families, and institutions change, and that they are composed of individuals whose interests and efforts toward a common cause change and are renegotiated. His history explains how there is not necessarily either tragedy or unity in managing common property, but rather more nuanced approaches that can work. The elements that have helped the community's enterprise to succeed include recognizing diversity while affording everyone the same benefits from the resource, and flexibly adapting to changing circumstances and to individual differences in interest and ability to contribute.

Oral histories like Martínez Rivera's are well suited to investigations of resilience and adaptation, which depend on both social and physical infrastructure. In 2015, Puerto Rico suffered a severe drought, and it became obvious to the entire population that potable water was a finite resource. The government water system established water conservation policies through water service interruptions starting in May,[17] and directives that included fines for washing cars or homes with pressure hoses beginning in June.[18] This resulted in hundreds of thousands of people losing access to running, potable water for stretches of several days at a time. The potable water that people did receive was of lower quality, due to the increased concentration of contaminants in the reservoirs.

These water conservation efforts may have come too late, since the reservoirs and river flow were already notably declining. By the time the drought ended in the La Plata River watershed, where the Doña Mayo community is located, the Aqueduct and Sewers Authority had spent more than $4 million clearing the sediment from lake reservoirs and dealing with the expenses of higher water treatment costs.[19]

Water rationing measures have important public health consequences, and any policies restricting water access are likely also to have political consequences. The government's response to the crisis was hampered by the fact that widespread water conservation measures are difficult to implement in large populations, where individual households may have differing perceptions of the needs of water conservation and where investing in a private cistern can allow households to continue using water as usual, even during enforced service interruptions.[20]

Small-scale community water systems faced similar challenges to those of the Aqueduct and Sewers Authority during the drought, but the interview with Martínez Rivera shows the way his community's system was prepared for such an extreme shortage. His narrative shows that the possibility of water shortages and the strategies for adapting to them are part of the institutional memory of the community water managers. Oral histories of community water system managers show that these systems can respond quickly to crises like drought: because of the close connections with the members of the community, these water managers can address their neighbors directly about the need to make adjustments in the water systems or in usage. Social pressure helps enforce adoption of water use restrictions, and the smaller size of the system and proximity to the water source allow the members of the community to directly observe the changes in stream flow or the level of their wells. This allows them to quickly adjust. Telephone interviews conducted during and immediately after the 2015 drought showed that most of the community water systems in the Health Department database were able to adjust to the shortage and had no problems with water access. This was in sharp contrast with neighborhoods that received water from the state system, which frequently went without water service for more than three consecutive days. Just as Martínez Rivera described while recalling other water shortages on the island, interviewees after the 2015 drought observed customers of the government water system coming to collect water from several of the community-managed systems.

Oral histories show that a sense of pride in community water systems is built from these systems' ability to help others access water in moments of need. A common refrain in the oral histories of community water managers is that while technocrats consider their systems primitive or rudimentary, they are able to provide water during environmental emergencies. The interviews recognize that the same characteristics that make the system "rudimentary" are also the characteristics that make them important for adaptation and conservation. For example, during the drought of 2015, it was estimated that more than half of the water produced by the government water systems was being lost somewhere between the water filtration plant and the households.[21] Managers of small community water systems, in contrast, often tell stories of noticing changes in the water throughput and quickly addressing them. Martínez Rivera's oral history demonstrates the kind of local knowledge needed to do this, encompassing both the water network and the technology used to place the tubing in the community. This kind of detailed knowledge of the water system is difficult to find in large governmental systems, where leaks can go undetected for years. While both small and large systems must confront infrastructure problems and waste, there may be useful lessons in his narrative and in similar success stories of other small systems about how to archive local knowledge about the water distribution system.

The small water system that Martínez Rivera describes is more flexible and better prepared to fix emergent problems, but it is also proactive in avoiding future risks. Because the community knows exactly where its water comes from, it has organized to conserve the forests within the watershed that feeds the water system. He knows the value of ecosystem services to water quality; his narratives mention how the community chose to locate the water pump in a forested area. Other oral histories include accounts of community members organizing groups to demand that people upriver do not alter the streams or contaminate the aquifer.

To my knowledge, the most striking case of local community environmentalism in Puerto Rico stemming from the need to protect the quality of their community-managed water occurred in the Guacio community. On multiple occasions this community successfully fought legal battles against their town's mayor and developers, eventually forcing the closure of a solid waste disposal site near their water source.[22] Martínez Rivera's recorded oral history does not contain such an extreme story of environmental activism, but it does demonstrate that his community has similar organizational resources, as well as the will and expertise to carry out legal battles to defend their water system. During conversations with Martínez Rivera while visiting the water pump, he pointed out a patch of deforested land that he had brought to the attention of the Puerto Rican Natural Resources Department in order to halt a plan to extend the deforestation. Other interviews with water system managers also reflect this interest and ability to conserve forested land for the benefit of community water: the stories include leaders preventing local housing developers or farmers from deforesting land in the watershed or establishing land-use activities that could ultimately affect the communal system's water quality. In this way, leaders of community water systems naturally become conservation stewards, without the intervention or education of a natural resource agency or outside environmental group.

It is said that "oral histories give voices to the unheard." Martínez Rivera's voice and those of other small community water systems are the unheard voices in the environmental movement in Puerto Rico. As in other places, the voices of environmentalism in Puerto Rico are more often dominated by groups and individuals focused on stopping large development projects or the contamination of natural resources. These influences are important in bringing the citizenry's attention to major environmental issues. However, the oral histories of the unheard people involved in community water systems are a reminder that other voices calling for attention to environmental issues are still present. Martínez Rivera's narrative is important, since it brings attention to an often-neglected fact: conservation and adaptation to environmental change is done by everyday people at household and local scales. It shows how people understand the role of forest ecosystems for water quality, without resorting to ecosystem service matrices or other frameworks by which scientists construct concepts of ecosystem services.

Oral history can do more than attract media attention toward the causes of unheard voices. These individuals, without public attention or media platforms, will probably keep working to support their environmental interests. However, the substance of their oral histories can remind us of the reasons *why* large, vocal, or empowered groups with abstract environmental goals do not carry forward all of the world's conservation efforts. When people and resources are intimately connected, they can support the same kinds of goals in a context beyond that of a polarized battle between humans and nature. Humans and water have a close and complex relationship, so it is important not to dehumanize water as we conduct the necessary process of evaluating it on a chemical or microbiological basis. Despite the pressures of more than fifty years of environmental and social change, it is possible for enterprises like the Doña Mayo community water system to survive and succeed.

Oral histories help us hear how water is valued in a human-centered way, and compel us to consider other disciplines and methodologies that could be brought to bear on questions of water quality and access. The purpose of recording and sharing oral histories like this one is not to undervalue the work of public utilities or of the chemical and biological measurements of water quality, but rather to serve as a reminder to consider the input provided by people who are local, who are experienced, and who have unique connections to infrastructure, agencies, and households. In this way, we can build a more effective picture of what "water quality" means, and a clearer view of its connections to other human and environmental needs. Martínez Rivera's narrative also resonates in a broader context: oral history is a field that has much to offer environmental research. Oral histories introduce new questions and shed light on dynamics that cannot be seen if we only rely on the scientific narrative. The collection of a community leader's memories, glimpsed through a recorded session focused on the life and experience with the land, provides images difficult to capture with other methods. Oral histories are especially useful to environmental and geographical research because they capture memories tied to space and experiences in the landscape. These memories situated in the landscape carry both environmental knowledge and social capital. The images they show us may be in conflict with the information extracted from statistics, graphs, and figures, but these should not be seen as conflicts challenging our duty as scientists but rather as a reminder that the environment is full of memories, and that people live with them. These memories can serve us as data, helping us to develop more inquiries and better strategies to share scientific findings about environmental processes with the public. They can serve us as an indicator, to help us detect prejudice in our paradigms or myths in people's shared understanding of nature. While the complexity of oral history will make our sometimes-needed scientific generalization more difficult to attain, it enriches our environmental research by making it more humane.

Postscript: As I was preparing this manuscript, Hurricanes Irma and Maria ravaged several Caribbean islands, including Puerto Rico. I interviewed leaders of small community water systems after the hurricane, many of whom described how their systems provided water to neighboring communities and were centers of emergency support, as during previous hurricanes. These conversations conveyed a heightened sense of the importance of these water systems. Other leaders, however, described the hardships associated with the total infrastructure collapse in Puerto Rico post-Maria: without electricity or access to fuel, their water systems were inoperable. In Doña Mayo, floods damaged the water pump and the generator, and fuel and parts were difficult to obtain. After attempting for more than two months to contact him, I finally spoke to Martínez Rivera in December 2017. Despite the evident difficulties, he expressed a fierce desire to defend and preserve the water system:

> Because it's been so many years of effort, and it's my legacy, and that of many people, to get to what we have today . . . It's been our blessing after

the hurricane. It's been the greatest blessing, because now I look at other sectors, where I go to work, at how people carry water, and they carry it in a way that makes you say "wow, we're so blessed in my community, we don't have electricity, but we have the most valuable thing." Even if it's off one day and on the next, we have water that when you turn on the tap of the faucet in your house, it gets to you.

Notes

1. EPA, "Drinking Water: EPA Needs to Take Additional Steps to Ensure Small Community Water Systems Designated as Serious Violators Achieve Compliance," US Environmental Protection Agency, March 22, 2016. https://www.epa.gov/sites/production/files/2016-03/documents/20160322-16-p-0108.pdf.
2. EPA, "Environmental Justice 1994 Annual Report: Focusing on Environmental Protection for All," US Environmental Protection Agency, 1995. https://nepis.epa.gov/Exe/ZyPURL.cgi?Dockey=40000AFB.txt
3. "Enforcement and Compliance History Online," US Environmental Protection Agency, https://www.epa.gov/laws-regulations/summary-safe-drinking-water-act).
4. Departamento de Salud de Puerto Rico, "Reporte de Brotes de Enfermedades Asociadas Al Agua de Puerto Rico 1976–1998," San Juan, Puerto Rico: Programa Agua Potable, Departamento de Salud de Puerto Rico, 1999.
5. "EPA Adds Corozal Well Site in Corozal, Puerto Rico to the Superfund List," US Environmental Protection Agency, March 13, 2012. https://archive.epa.gov/epapages/newsroom_archive/newsreleases/7346cebd7aed2918852579c0005f8dcb.html
6. Sara Del Valle Hernández, "Sequía Afecta Calidad Del Agua Potable Servida En El Área Metropolitana," *El Nuevo Dia*, February 9, 2017, http://www.elnuevodia.com/noticias/locales/nota/sequiaafectacalidaddelaguapotableservidaenelareametropolitana-2289454/.
7. My focus on individual memories is in no way a call to undervalue either the importance of scientific measurement or the reach of SDWA in protecting both public and environmental health. The SDWA has been crucial for protecting communities at risk of health impacts: for example, the application of SDWA in the case of the Flint city water system in Michigan in the United States highlights the importance of this law in protecting the health of disenfranchised social groups.
8. In part, this race to disconnect community water systems is rooted in misunderstandings about appropriate measures of tropical water quality which add to the perceived notion of risky water, something I discuss elsewhere. See Javier Arce-Nazario, "The Science and Politics of Water Quality," in *The Palgrave Handbook of Critical Physical Geography*, ed. Rebecca Lave, Christine Biermann, and Stuart N. Lane. (London: Palgrave Macmillan, 2018), 465–83.
9. "Principles of Environmental Justice," 1991, last modified April 6, 1996, http://www.ejnet.org/ej/principles.html.

10. Gerardo Millán-Ramos and Jorge Martínez-Ramírez, "Integration of Public and Private Resources at the Solution of a Public Health Problem: The PPW Experience in Puerto Rico," (paper presented at the annual meeting for the Inter-American Association of Sanitary and Environmental Engineering Caribbean Regional Congress, San Juan, Puerto Rico, 1997), 2.
11. Garret Hardin, "The Tragedy of the Commons," *Science* 162 (1968): 1243–48.
12. E. Schlager, and E. López-Gunn, "Collective Systems for Water Management: Is the Tragedy of the Commons a Myth," in *Water Crisis: Myth or Reality*, ed. Peter Rogers et al. (London: Taylor & Francis, 2006), 43–58.
13. Karen Bakker, "Neoliberal versus Postneoliberal Water: Geographies of Privatization and Resistance," *Annals of the Association of American Geographers* 103 (2013): 253–60.
14. Vandana Shiva, *Water Wars: Privatization, Pollution, and Profit*. (Berkeley: North Atlantic Books, 2016).
15. Elinor Ostrom, *Governing the Commons: The Evolution of Institutions for Collective Action* (Cambridge: Cambridge University Press, 1990).
16. Esther Wiesenfeld, "The Concept of 'We': A Community Social Psychology Myth?," *Journal of Community Psychology* 24 (1996): 337–46.
17. Estado Libre Asociado de Puerto Rico, "Orden ejecutiva para declarar un estado de emergencia repecto al uso de las aguas por sequía y establecer un plan de acción interagencial," OE-2015-011 § (2015).
18. Autoridad de Acueductos y Alcantarillados, "Orden administrativa para regular el uso de agua potable durante la vigencia del estado de emergencia por sequía," OA-2015-02 § (2015).
19. Keyla López Alicea, "Lo Que Dejó La Sequía Al País," *El Nuevo Dia*, January 14, 2016, http://www.elnuevodia.com/noticias/locales/nota/loquedejolasequiaalpais-2150439/.
20. Francisco Rodríguez-Burns, "AAA Aumentaría Tarifa Si Hay Uso Desmedido Del Agua Potable," *El Nuevo Dia*, August 6, 2015, http://www.elnuevodia.com/noticias/locales/nota/aaaaumentariatarifasihayusodesmedidodelaguapotable-2083202/.
21. Aurora Rivera-Arguinzoni, "AAA Pierde Lo Que Invierte." *El Nuevo Dia*, August 5, 2015, http://www.elnuevodia.com/noticias/locales/nota/aaapierdeloqueinvierte-2082494/.
22. Helen Santiago, "Acueducto Rural Guacio: Algo Más Que Un Acueducto." *Maguey*, September 2010.

8

Environmental Guardians

Learning from Māori Perspectives on Geothermal Fields

Caren Fox

Oral history by Caren Fox with Aroha Campbell at the Ohaaki Power Station near Reporoa, New Zealand, January 14, 2017

In 2008, I served as presiding officer for the inquiry and report by the Waitangi Tribunal in New Zealand, He Maunga Rongo: Report on Central North Island Claims.[1] *The Central North Island region has the highest number of geothermal fields in the world, with seventeen in the Taupō Volcanic Zone (TVZ) of the Central North Island.*[2]

Māori people have had an ancient relationship with this resource since they arrived in Aotearoa/New Zealand from their traditional homeland of Hawaiiki in the Polynesia triangle. Following the colonization of Aotearoa/New Zealand by the British, this relationship was not recognized in law and policy, including during the development of certain geothermal fields of the TVZ through the years 1950–1991. In her interview, Aroha Campbell, from Ohaaki near Reporoa, a small village south of Rotorua in New Zealand, explores the impact of the New Zealand government's failing to provide for this relationship and the resulting unsustainable exploitation of their geothermal resource. Ohaaki was commissioned in 1989 by the Power Development Division of the New Zealand Electricity Department of the New Zealand government to increase the supply of electricity to the national grid. When the power station was constructed, it could generate up to 104 megawatts (MW), but the unsustainable exploitation of the geothermal resource led to the gradual depeletion of sections of the geothermal field. This has meant that the maximum net capacity generated by the Power Station has since dropped to 65MW. The decline has been attributed to a number of factors related to the operation of the power station, but the exploitation of the shallow acquifers and the initial failure to reinject processed

Caren Fox, *Environmental Guardians* In: *Oral History and the Environment.* Edited by Stephen M. Sloan and Mark Cave, Oxford University Press. © Oxford University Press 2022. DOI: 10.1093/oso/9780190684969.003.0009 *geothermal or other fluid after extraction have been identified as major contributing causes.*

Subsidence of the land has occurred, resulting in the land near the plant being lowered to such an extent that it is now submerged by water from the nearby Waikato River. The impact of the subsidence has affected the road used to access village utilities and sacred sites, leading to the construction of a new road and bridge and remedial work on those utilities and sacred sites. The effects of that subsidence on the local Māori community and the arrangements they came to with the power station owners to overcome any deterimental effects are discussed by Aroha Campbell.

Campbell has been a driving force behind the remedial work completed at the Ohaaki site for her sub-tribe Ngāti Tahu. She is one of a number of trustees behind the successful geothermal operations of a number of her tribal authorities. These include the Ngāti Tahu Tribal Trust and the Tauhara North No.2 Māori Land Trust (located near the town of Taupō, New Zealand).[3] *Campbell has also been a director on the boards of several New Zealand companies with interests in geothermal development, including Tauhara Development Limited, Rotokawa Generation Limited, and Rotokawa Joint Venture Limited. At the time of the interview, she was the chief executive of the Tauhara North No.2 Trust Māori Land Trust (Taupō).*

CAMPBELL: I was born 1958 in Rotorua and lived all my life in Rotorua. I am the youngest of three with three younger *whāngai* [customarily adopted children]. My interest in the land came from my dad because he used to go to a lot of the land meetings in Reporoa. And so, growing up and at primary school I used to travel alot with him to all of those land meetings. I never went to University—my education is as far as secondary school and short courses.

Growing up I actually didn't like coming to Reporoa because it was quite a long way—for a little kid, it seemed like this long journey, the road conditions were lot different from today, and then there was just too much forestry around the place and big huge trees.

Certainly at Ohaaki it never seemed to be a friendly place with the bush, geothermal and how loud it can be. The *ngāwha* at Ohaaki was pristine back then. I had no fear in terms of walking right up to the *ngāwha* edge—seeing the steam rising and knowing don't dip your finger or toe in the *ngāwha* because it's far too hot. It serviced the bathing pool—for bathing and further down washing clothes. Back then, those times were very simple and that is what I saw growing up.

Then in 1989, after discussions with a cousin of mine, Nuku Hurihanganui—he persuaded me to put my hand up and he nominated me to be a responsible trustee on Ngāti Tahu Tribal Trust.

The tribal trust was formed back then. We did have *whānau* [family] that wanted the trust to change its name—so I, along with my fellow trustees, made it clear that we are only responsible for six blocks of land in and around the Ohaaki Power Station. I am the last remaining trustee of those appointed in 1989. I am also the last original trustee on Tauhara North No.2 Trust appointed in 1992. I've always been mindful not to spread myself thin across too many boards, that's why I only serve within Ngāti Tahu. And so that is my journey in relation to the land, it's been a slow, long, long journey and there is still a long long way to go.

At Ohaaki, we maintain six blocks of Māori land. The Ngāti Tahu Tribal Trust is one entity. Then we have *Ohaaki Marae* [traditional meeting place] being called after our eponymous ancestor Tahu Matua—that certainly carries, for me, a large responsibility. Since becoming a responsible trustee of the Ngāti Tahu Tribal Trust, it's been more about reacting to situations because of the lease agreement of the Ohaaki lands to the government. In hindsight the lease didn't provide Ngāti Tahu with an opportunity to commercialize or to participate in the development—or in understanding the resevoir and or even providing cultural advice. So we just sat there passive and really supporting the Ohaaki Marae Reservation Trust. It was more because the Marae Reservation Trust didn't have any resources at all.

With the commissioning of a power station, a community percentage or fund goes back to the community. For the construction of this power station, Ngāti Tahu and Ohaaki Marae Reservation got zero. That community fund went to build the squash center in Reporoa and some other things—it supported things around the Reporoa community, but nothing for the *marae*—nothing.

So this power station feeds the national electricity grid. There has always been a perception within Ngāti Tahu owners that Ngāti Tahu Tribal Trust was benefiting from the output of this power station. And that certainly is not the case. We received a one-off payment, and we have been very frugal. That was approximately 1999—after the memorandum of understanding and finalizing the lease agreement, and it is still in place today.

As a trustee I certainly saw the *mamae* [sorrow and pain] caused from the geothermal development—and more so with the *ngāwha* fluids disappearing and the work around to resolve the issue.

Seeing the changes with the *marae* [meeting place] as well as with the river levels has been sad. That is because of the subsidence around the marae—something like 3.7 meters.

There are shallow, medium, and deep aquifers within the Ohaaki [geothermal] reservoir. And so fluid was actually taken from the shallow aquifers and not returned. That has resulted in subsidence of the land. What I recall of the river edge is that it was halfway out in the current middle of the river. And looking at the old photos—the *marae* [traditional meeting place and buildings] used to be sitting up high, and then you used to walk down to the river edge, well you don't do that anymore it's wiped that out. So that is what I recall growing up.

. . . And so that's a deep hurt for *whānau* [family] and it still remains today. And I think that that has been a real barrier for some of our *whānau* to move on. I guess it is laying blame with, certainly with the government and then with Contact Energy.

In the 1980s the government's New Zealand Electricity Department privatized their interests including Ohaaki Power Station into a state-owned enterprise called ECNZ. In 1995 and 1996, ECNZ split and formed a second electricity state-owned enterprise, Contact Energy. Contact Energy was sold in 1999 into private ownership, and for a period from 2004 its major shareholder was Australian-based Origin Energy. Campbell explained the relationship with Contact Energy.

CAMPBELL: It has been an interesting one in terms of the culture of a commercial entity. Back then, it was very hard to establish and build a relationship with the owners from Australia— very hard. So it has taken a very long time to build the relationship with the senior management. So one could say we woke up, took a deep breath and looked at what we could do to gain some leverage, and that was prior to 2011, the reconsenting for this power station. That was because they had to go through a Resource Management Act 1991 (RMA) process.[4]

They applied for a thirty-five-year consent to carry on operating the power station. We went through a consultation process and Contact Energy were really good and supportive. I believe that they were finally listening to us. We are known as the Land Owner Collective. There are three groups of land owners—Ngāti Tahui Tribal Trust, the Ohaaki Marae Reservation Trust, and Tahorakuri A1 section 30 Block. The last block is next door to the *marae* [meeting place].

So the three got together, as this was our one-time shot at it. Even though we are all very closely interlinked, we have been really separated in our own little silos and this was our one time to come together and work collaboratively. Not against Contact Energy—but with Contact Energy to gain participation in the monitoring of the Ohaaki field, the cultural monitoring of existing work programs, to work through the *mamae* [sorrow and pain].

So with our lawyers, we drafted up an agreement in principle and that was signed back in 2012. It clearly identified our collective purpose, our objectives, and how we would work together. And then out of that, those three entities, we identified three women to negotiate with Contact Energy. So that was myself, Dotty Raroa for the Marae Reservation, and Tess Kora for A1 section 30 Block.

Contact Energy was really good in terms of fronting up, coming out to the *marae* [meeting place] and meeting with the different trusts and, knowing that we didn't have any resources, providing that support and resource.

In September 2013 we had the consent hearing, and the night before the RMA consent hearing our lawyers presented the agreement litigation package. That was approved by the three trusts, and it was signed at Ohaaki on the night before the RMA hearing.

The package includes "cultural monitoring," we never had that before. So that concerns our sensitive areas, including the *marae*, *ngāwha*, the fertility rock, and the *urupā* [burial ground] areas that we have identified. Whilst they can still traverse over the land to do the usual monitoring, they can't unearth for anything else than the usual operations in there. So we know what those operations are. If it is anything outside that, they have to then come back to us or it is a no go.

That was a huge achievement to include cultural monitoring, as that allows participation and, to a certain degree, control. The company is also building a bund to protect the *marae* [meeting place]. We are monitoring and have input to the bund—the *marae* protection bund—we've had a full time monitor in that space. Contact Energy is paying for that at around about $3.2 million. What had happened there is that from Contact Energy's perspective it was just the protection of the *marae* [meeting place] as per the resource consent. But the Ohaaki Marae Reservation Trust took a different position and said they can't just protect the *marae* [meeting place] area knowing that there is land on both sides which is Ngāti Tahu and A1 section 30 Block that was inundated. What they did, with the owners' approval in August 2013, was a plan to connect the road from the *marae* [meeting place] to Ohaaki Road, as it was under water as well as the A1 section 30 Block.

So with the bund development we also ensured that once it was completed that the Marae Reservation will have $700,000 to allow upgrading of the *marae* buildings. So coming from zero dollars to protect the *marae*—then they come away with $700,000—that's a good starting point.

Contact Energy has been monitoring stations around the *marae* and along the river edge as well. Those have been in there since the early 2000s. What they have is good data, and the scientists have indicated that the subsidence is slowing down, but there is a likelihood of losing another 30 millimeters.[5] Within our mitigation package we have the level at 29 millimeters, then we have the 30 millimeter buffer, and if it goes one millimeter over that, then we are back around the table talking with Contact Energy. So we had two options within the mitigation package: (a) The first was keeping the *marae* in its current location and its protection, we need to have a future proof in place should these predictions go sideways and we do have flooding, what happens then? So option (b) putting in measures for returning back around the table to discuss removing the *marae*.

Subsidence normally means that the field itself is starting to become depleted. It happened here because their production zone was shallow. In geothermal fields there are three aquifers in the field. They hadn't actually drilled down to the deep aquifers. So the fluid from the shallow aquifer was extracted and not returned effectively, and that is why subsidence has occured.[6]

The company is now reinjecting in the right place. If reinjection in the right place had taken place from the start, the field would not have collapsed, the subsidence would have been millimeters not meters. They are now reinjecting back into the middle and deeper aquifers. They would have started that in approximately 2012. So Contact Energy are actually drilling deeper now. Now that is the lesson learned for Tauhara.

With the *ngāwha* [geothermal pool], the *ngāwha* has been filled with reinjection fluid and that has its own challenges. We removed the old bath house last year. That is because within the mitigation package there is a financial contribution for a new bath house.

The water for the bath house will come from the *ngāwha*. It will be batheable. With the *mamae* [sorrow and pain] from our *whānau* [family]—it is because they remember clear pristine water, that's what they wanted to see. So for the last three years with Contact Energy we have been trying to find a solution for that. And there have been different alternatives we have considered.

So the mitigation package has allowed the three land trust entities to have financial funding to do certain things. So each of the groups had a package each, and we are three years on and the fund is still intact. This meeting and all of this furniture in this room, has been furnished by Contact Energy for us to use as we see fit. It also included computers, screens, and we passed them on to the *Kōhanga* [early childhood center], the *marae*, and *whānau*.

Ngāti Tahu Tribal Trust is in development mode, finally having the ability to look differently at our potential and our opportunities for commercial development. Our vision is to reestablish the Ohaaki community and we have been on that journey for several years. And certainly with Contact Energy, their support has been phenomenal.

As a result of the changes, Contact Energy had to change their thinking and processing here and at Wairākei as well. I think it was an attitude change for the organization as a whole. And being driven by the board down. I think from the shop floor to middle management I witnessed their passion, values and strong connection, and they shared our *mamae* [sorrow and pain] around here. And so I think that it has been a real change for the betterment for Contact Energy.

The power station was originally built for 120MW. It was the predevelopment drilling that actually sucked out all the initial shallow aquifer. So this plant has been only producing between 40 and 60MW.

In the early days as a guest speaker at a conference, a question was asked about the comparison between the two organizations—Mighty River Power, now Mercury, and Contact Energy—I used to say there was no comparison. Mighty River Power won hands down. They started early and recognized the value in terms of the relationships with Māori and they started making changes very early in the piece, when Contact Energy was still yet to start.

So for Ngāti Tahu Tribal Trust, the resource consent was our leverage. Not to be demanding, but saying that this is what is right and how do we work together to make it happen. Now we have a really good Land Owner Collective relationship. We have a management team that interacts with Contact Energy.

Campbell is also a trustee on the Māori Land Trust Tauhara North No.2 Māori land block—situated south of Ohaaki on the Rotokawa geothermal field. She proceeded to explain the difference between the business models of Ohaaki and Rotokawa.

CAMPBELL: In the begining for Tauhara North No.2 Trust, it wasn't a pretty block because there is actually no physical sign of geothermal on our block. Compare that to what one can see at Ohaaki—there are fumaroles, there is a *ngāwha* [geothermal pool]. At Rotokawa you do have the Rotokawa Lake. There are some visible areas on the Parariki land block, but certainly on Tauhara North No.2 block there isn't anything at all, nothing. When I traversed the block in the early days my immediate thought was, This is the ugliest piece of land—no wonder nobody wanted it, because they couldn't see any value to it, but the real value was underneath in terms of the geothermal itself.

We have always had participation in the development and management of the reservoir including the Rotokawa power stations—unlike Ohaaki. The lessons learnt and difference is, at Tauhara North No.2 Trust, you've got one block of land, one set of owners, and one organization to deal with. That in itself is a game changer.

At Tauhara North No.2 Trust we have had a lot to be thankful for in terms of the experiences at Ohaaki for Ngāti Tahu Tribal Trust. At Rotokawa, we acknowledge what happened at Ohaaki—and we were determined to ensure we had control of our destiny, participation in all areas of the development including the environment. When we first started, subsidence was a number one priority on the list. It is still a priority, but it's just a component now in terms of monitoring and risk.

The Tauhara North No.2 Trust partnered with our first partner the Taupō District Council, they decided it wasn't their core business so they sold that interest to Power NZ. And then built the 18MW power station. This is where we took advantage of the situation—learning and upskilling ourselves.

During the state-owned enterprise sales [in the 1980s], Power NZ sold its interests at Rotokawa to an Australian company Transalta—we agreed for the transaction to take place with a $2 million ticket, and also included a clause to give us a twelve months opportunity to purchase the station outright.

During that twelve months we initiated a tender process. And we chose Mighty River Power as our partner. So they became the hundred percent owner in the Rotokawa Power Station.

The planning for the station started in 1994, and that was commissioned in 1997.[7] We had representation on that [governing] board even though we didn't have an equity stake. So Mighty River Power was very caring in terms of supporting and building our capacity and our knowledge of the resevoir. So we had a seat on that board. However, from 1997 through to 2004, we built up our knowledge of the reservoir and operations of the power station. We wanted to make sure that we didn't end up like Ohaaki.

The power station at Rotokawa operated for about thirteen years generating approximately 34MW—and then we decided that we wanted to take on a risk profile, we wanted to have equity. I wasn't on the commercial board then. I didn't come on until 2006. I did go to a couple of the meetings with the chair at the time, Phillip Sullivan, and our advisor, Buddy Nikora. The overall operations at that point certainly gave the trustees the confidence that we were on the right track—we didn't know everything, but we were certainly comfortable that it was timely for us to move into an area of taking on a risk profile. We did so listening to our owners and their aspirations. And there was no way that we were going to have an oversized power station on the land or allow the reservoir to be depleted or not managed correctly—for the benefit of our owners and descendants.

That is when the planning for Rotokawa Stage II began. Within New Zealand, due to the reduction in natural gas, geothermal became the "in" thing again. So for us it was a waiting game. And we just happened to be there at the right time and the right place and we made the right decision to build a power station.

Ngā Awa Purua [power station] was the result of this planning, and it was originally set for a 80MW power station. We thought it's a bit bigger than the 34MW, but from what we learnt about the reservoir we knew that it had the capacity to maintain both. For Tauhara North 2, we chose who we wanted as our partner and that was Mighty River Power, we built and strengthened the relationship.[8]

We decided we wanted to start the second project of Ngā Awa Purua and initiated discussions with Mighty River Power. They were good about ensuring that they were always there to give us information to get us going on our journey.

In 2010, Ngā Awa Purua was commissioned and our relationship has grown stronger. So it's just gone from strength to strength at all levels within their organization. Ngā Awa Purua Power Station was promoted as an 80MW power station. But through advanced technology with Fuji and Sumitomo they came up with a triple flash single shaft turbine.[9] In 2007, at Kawerau, Mighty River Power had built a 90MW power station with a dual flash. With the proposal for a triple flash—this meant that we could have three bites at the cherry, at the energy, for the same price. So that 80MW power station became a 132MW.

Both stations at Rotokawa feed the national grid. We have worked very closely with Mighty River Power, and participated within all facets of the resource consent. So thankfully with the support of those original trustees we put a stake in the ground in terms of saying who we are, the land owners of Ngāti Tahu, and we acknowledge the late Rāwiri Te Whare. And I guess that was good for the Trust to keep us focused on our end goal and not get caught up in Māori politics. So that is why we kept it simple, and also flying under the radar, we fronted the [public] consultations side by side with Mighty River Power. And we did our share of the *mahi* and stood on our own. It was a great learning platform for me.

For me, I needed to understand where they were coming from, what was their drivers, and for them to understand what was important to us. We always maintained that they had all the expertise and the capacity and resources, we are just a small land trust.

We agreed to go down this journey for a new power station. So when there is an issue that's on the table, don't come to us looking for a solution that only satisfies your needs. You come to the table with a solution that satisfies your needs and satisfies ours, and then we will get on very well. We made it clear that in their planning, including engagement with the Trust, they needed to put in extra time, and extra resources. And the fruits were realized very early in the project.

Now we also have to consider the Ngātamariki Field northeast of Taupō, the hottest geothermal field in this district. Again we have worked with Mighty River Power, and Ngātamariki power station was commissioned in 2013 by Mercury Energy [previously known as Mighty River Power].[10]

The Tauhara North No.2 Trust's concerns over subsidence have been addressed, as Mighty River Power use state of the art reinjection technology. This is important as a business model, as it is expensive and the deeper the wells—the greater the cost.

[Thus the values that Māori bring with kaitiakitanga *[guardianship/stewardship] have influenced the relationship the Tauhara North No, 2 Trust has with Mighty River Power.]*

Campbell continued:

CAMPBELL: Scientists will say that there are no links between the geothermal fields, but I always say if you go down far enough you hit magma and it is all linked. That's why we care for the resource.

There are many stories relating to the ancestor Ngātoroirangi in this district. I remember a conversation with Rāwiri Te Whare about Ngātoroirangi. He said that Tahumatua, our ancestor, was actually here before Ngātoroirangi so, therefore, Tahumatua would have already found our Ngāti Tahu lands. And so I said to him at the time—we can still acknowledge Ngātoroirangi, but ownership is with Tahumatua and his descendants.

So that's why with Tauhara North No.2 Trust—our logo is an acknowledgment to Ngātoroirangi and to the *maunga* Tongariro, but that is all it is—an acknowledgment. Tahumatua is the eponymous *tūpuna* of Ngāti Tahu, he resides at our instutional *marae* at Ohaaki. We are the *kaitiaki* [guardians].

Commentary

Understanding the relationship of Campbell and her people to geothermal resources means understanding how ancient the history of living with the geothermal activity is. It also means understanding what it means to them. Tradition and oral history provide some of the answers, as do the eyewitness accounts of early visitors from Europe.

In 1859, the Austrian geologist Ferdinand von Hochstetter, for example, carried out field explorations in the Central North Island. This was long before Western scientific geological knowledge of the TVZ existed. He visited Lake Taupō and had extended conversations with Iwikau Te Heuheu Tukino III,[11] the paramount chief of the tribal nation Ngāti Tūwharetoa, the tribe located on the shores of Lake Taupō. As the Waitangi Tribunal noted in its report, Hochstetter observed, after his interview with Te Heuheu, that "the natives have very correctly brought the hot springs directly in connection with the still active volcanoes, though they have clothed their conceptions in the garb of legend." He later noted that the legend of Ngātoiroirangi "affords a remarkable instance of the accurate observation of the natives, who have thus indicated the true line of the chief volcanic action upon the North Island" of New Zealand.[12] Not surprisingly, he also found Māori settlements alongside many geothermal surface features within the TVZ.

The relationship of the tribes of the Central North Island to geothermal features is associated with their respective settlement patterns and with the *tohunga* [navigator and priest] Ngātoroirangi and his journey throughout the region. That journey has been described by Chris Winitana of the Ngāti Tūwharetoa tribe:

> When [our ancestor Ngātoroirangi] arrived at these parts he ascended Tongariro. He reached the summit and was overcome by a snow blizzard. Knowing that he would most certainly perish in the storm, he invoked his ancestors Te Pupū and Te Hōata, the elders of the fire clan of Hine-tapeka, to come to his aid. He implored Kautetetū to produce the fire born of friction that he needed to save his life. He called to his sisters Kuiwai and Haungaroa, who were still in the homeland Hawaiiki, and they sent their fire ancestors to help their brother. Te Pupū and Te Hōata travelled underground with their precious gift and at different places along their route emerged to ensure they were travelling in the right direction. These places became geothermal or volcanic spots and include Whakaari, Tarawera, Paeroa, Orākei-korakō, Wairākei, Taupō, Tūaropaki and Tokaanu. The fire emerged at the summit of Tongariro and the old priest was saved.[13]

The relationship enjoyed by the tribes of the Central North Island with their geothermal resources has continued to be both a physical and metaphysical one. The late Hāmuera Mitchell (Senior), elder and chief of the Te Arawa tribe, whose sub-tribe is Ngāti Whakaue and who was from the village of Ohinemutu, on the shores of Lake Rotorua, described the deep significance of the relationship in 1993. His translated statement, published in part by the Waitangi Tribunal, records his right to represent his tribe. He then described the geothermal features around Lake Rotorua, which is approximately 78.8 square kilometers and the second largest lake in New Zealand. He stated:

> That's how Ngāti Whakaue was in those days, and this place is 100% geothermal. All along the lakeshore was geothermal . . . we would go fishing for morihana . . ., we had no clothes (no *pakehas* [white people] around in those days aye), and we would catch a few and tie them on a string, we would be shivering then we would dig a hole in the *mea* [thing] by the lake and the water come up and we'd be warm. If we wanted to cook our fish straight away, we'd dig another hole, and cook our breakfast. That's how we were in the old days, we lived it we spoke it we sang it and we slept in it. Right over here, that's why it is called to Papaiouru. A *Papa* is a slab of stone, *Ouru* is a *ngāwha* [hot pool]— that's one boiling over there. That's why they called it Te Papaiouru. We used to sleep right in front here where it was warm, we would put our mattress there in those days, it was warm, the whole place was warm. I'll take you around quickly after we finished here, you'll have to leave your shoes on, otherwise you'll scald your feet. Now we regarded Ohinemutu specially, with great faith and sincerity that we treated our pieces of land and of course our *ngāwhas,* with decorum . . ., we even treated them like persons at times. We would speak to them, quite often the *ngāwha* would change in the degree of heat, it would be quite cold, so we speak to them, and say *'ata e koro, hei tīmata koe i tēnei ata'* [good morning elder are you going to perform this morning] and it would come back and say *'hei anō apopo koe ka hoki'* [when you return tomorrow maybe]. I can hear Bishop Kingi, he's always talking to them every morning. . . . But we've got hot pools all over the place at Ohinemutu. That's how we feel about them, that's how we feel about them today. . . .
>
> The *ngāwhas* come and they go. They're very temperamental. We look after them, we would be worried if they didn't come back. We used them when I was a child, for health purposes, cleansing, cooking. We use them today, everyday. We have songs, *waiatas* of our *ngāwhas* here. . . .[14]

Mitchell's village is one of several with geothermal features. Aside from geothermal springs in such villages, there are many other features such as fumaroles, geysers, and mud pits to be found in the region, along with the seventeen hot geothermal fields. These are: Kawerau, Tāheke-Tikitere,

Rotorua, Horohoro, Waiotapu/Waikite/Waimangu, Reporua, Ohaaki, Ngātamariki, Rotokawa, Orākei-Korakō, Wairākei-Tauhara, Horomatangi, Tokaanu-Waihi-Hipaua, and Mōkai-Tuarōpaki.

Māori in the Central North Island have claimed that these geothermal sites of the TVZ are one resource, albeit with three aspects to the TVZ. These are the surface features, the geothermal water or fluids, and the geothermal heat and energy located in the geothermal fields, and the subterranean resource.

Māori consider the geothermal resources of the TVZ to be part of an indivisible metaphysical system, one that is personified by the various elemental gods recognized in Māori cosmology. This Māori personification of the environment is gaining recognition in New Zealand. Tribes in other parts of the country, for example, have negotiated recognition for important environmental features in their territories, including sacred mountains, rivers, and lakes. Legislation for features such as the Waikato River recognizes these elements by giving them legal personality. The Waikato-Tainui Raupatu Claims (Waikato River) Settlement Act 2010, for example, declares in Part 2 that the Crown (the government) recognizes the statement of significance of the Waikato River to the Waikato-Tainui tribe. In that statement, the tribe declares the River to be their "... ancestor with *mana* (spiritual authority and power) and in turn it represents the *mana* and *mauri* (life force) of Waikato-Tainui." Most of the Waikato Hydro Scheme operated by Mercury (once Mighty River Power) is affected by this legislation.

Those developments have been taken one step further with the Urewera National Park. Under the Te Urewera Act 2014, the legislation recognizes the Te Urewera as a legal entity with all the rights, powers, duties, and liabilities of a legal person, whose rights are exercised by the Urewera Board, made up of tribal members and ministerial appointees.

In March 2017, the Te Awa Tupua (Whanganui River Claims Settlement) Act was also enacted by the New Zealand parliament. The act describes Te Awa Tupua as "the indivisible and living whole, comprising the Whanganui River from the mountains to the sea incorporating all its physical and metaphysical elements." The river is now a legal person and has all the rights, powers, duties, and liabilities of a legal person, for whom an entity called Te Pou Tupua take responsibility. That entity is composed of representatives of the tribes with interests in the river and of the New Zealand government.[15]

Māori of the Central North Island told the Waitangi Tribunal they wanted the same acknowledgment for their relationship with the geothermal features of the TVZ. They claim to be the *kaitiaki* (guardians) of the resource. They have used it for bathing, cooking, and heating since ancient times. In using this resource, they developed traditions, oral histories, and cultural practices to manage it. Collectively the tribes consider the TVZ to be a *taonga* (treasure), and the sub-tribes claim a right to control and manage it where located on their lands. The Ngāti Tahu tribe, of which Aroha Campbell is a member, have traditional relationships with the Orākei-Korako, Wairākei-Tauhara, the Rotokawa, Reporoa, Ngātamariki, and Ohaaki fields. All these fields fall within their tribal territory.

As Chris Tamihana Winitana of Ngāti Tūwharetoa tribe explained to the Waitangi Tribunal, Māori oral history records that the TVZ geothermal resource was created and transported to New Zealand. The story involves the separation of the primal parents *Ranginui* (Sky Father) and *Papatūānuku* (the Earth Mother) by the actions of their children—the various elemental gods. He explained that Māori believe in gods such as *Rakahore* (the father of bedrock and stone), *Pūtoto* (magma and lava), and *Parawhenuamea* (water). Therefore, the various gods and the ancestors, particularly Ngātoroirangi, assume a central role in the management narrative of the Māori landowners and the tribes of the Central North Island.

The Ngātoroirangi story, in particular, is repeated by the tribes of Te Arawa and Ngāti Tūwharetoa and their sub-tribes, although details and the manner in which they are acknowledged or presented may vary, as in Campbell's narrative about Tahumatua, their epoynomous ancestor. The places identified in the Ngātoroirangi story are all landmarks that identify geothermal features of the TVZ. The tribes claim that this indicates they have known of and lived with the geothermal resource since time immemorial.

They have described their knowledge and customary use of the region from White Island (Whakaari) in the north of the region through to Tokaanu on the southern shores of Lake Taupō and on to the three volcanic mountains—Ruapehu, Ngāuruhoe, and Tongariro. They have named geothermal surface manifestations and places of significance. These geothermal places include Horomatangi—who is the water guardian of Lake Taupō. According to oral history, his lair is in the spot where a new hot pristine geothermal field was found toward the end of last century. Horomatangi features in the Ngātoroirangi story when geothermal fire arrived from the Pacific homeland—Hawaiiki.

So, it is clear that Māori claim to have a significant relationship with the geothermal resource of the TVZ. However, that relationship is affected by New Zealand law. First, the common law recognizes that the rights and interests in geothermal resources are vested in the owners of the land under which a field is situated. They are entitled to the benefits of the resource when they seek to develop it. However, the government has appropriated the right to manage the resource and delegated that responsibility to regional councils. Thus, the management of geothermal resources is governed by the Resource Management Act 1991 (RMA).

In the Central North Island region, the Bay of Plenty Regional Council and Environment Waikato Regional Council manage the TVZ geothermal resource, including the fields. Importantly, their management plans refer to the need to manage geothermal developments wisely and sustainably, in consultation with Māori landowners and tribes and after taking into account or having regard to Māori values or *tikanga* (Māori customary law) including *kaitiakitanga* (guardianship/stewardship for future generations).[16] This is because the RMA requires that these matters be considered. It declares that all those exercising powers and functions under the RMA, including regional councils and the Environment Court, must recognize and provide for the relationships of Māori and their culture and traditions with their ancestral lands, waters, sites, sacred places, and other *taonga*, and having regard to their *kaitiakitanga* and the principles of the Treaty of Waitangi.[17] Thus, all those who have sought resource consents or renewed consents for development, including tourism or electricity generation, must also have regard to these matters prior to making an application for a consent.

Although these provisions in the RMA accommodate an element of the Māori interest in the TVZ, the tribes of the Central North Island assert that the RMA does not go far enough. They contend that they have the right to control and manage the geothermal resource in accordance with their own cultural preferences. Such preferences include elevating sustainability and prioritizing preserving their geothermal features for future generations. Before the Waitangi Tribunal, they claimed that the Treaty of Waitangi preserved to them the right to exercise control, guardianship, and management responsibility over the resource. The Waitangi Tribunal agreed with them, and it recommended to the New Zealand government that it should acknowledge and provide for Māori rights to manage activity within the TVZ with the Bay of Plenty Regional Council and Environment Waikato Regional Council. It also considered that the tribes had the right as landowners to act as the *kaitiaki* (guardians) of the resource where it manifests on their land.[18]

The issue of the extent of Māori rights is still subject to further litigation. What I am interested in is how Māori continue to be involved in the use and development of geothermal resources and what their values add to the management of these non-renewable resources in contemporary times.

Residential use and the day-to-day management of village geothermal surface features continues today in most Māori communities in accordance with the *tikanga* (customary laws and values) of the various tribes and sub-tribes. A visit to Whakarewarewa or Ohinemutu in Rotorua would demonstrate that they still use geothermal features daily for cooking and bathing, and the evidence given to the Waitangi Tribunal by the late Hāmuera Mitchell (Senior) is a clear example of the values they ascribe to the resource and how they manage their relationship with their geothermal features.

Secondly, the tribes have participated in tourism. Commencing in the nineteenth century, Māori have played a part in the development of the industry. Initially they acted as guides, running their own operations by taking tourists to famous geothermal features. Then, as colonialism took root, they themselves became features of the tourism product. Rarely, however, did they control the tourism trade after the 1900s. In recent times, more Māori operators are entering the tourism market, but the share in that market could be improved.

Aside from tourism, the other commercial use has been electricity generation. Kawerau was the first geothermal production field in New Zealand in 1956. It was a part of the Tasman Pulp and Paper Mill development[19] and has been extensively developed since then. According to the Bay of Plenty Regional Council's website, several companies now have consent to draw on the resource, including Mighty River Power (now Mercury Energy), Ngāti Tūwharetoa Geothermal Assets, Eastland Developments, and Te Ahi o Māui.[20] Wairākei became the first geothermal power station in New Zealand in 1958 and the second in the world. Successive New Zealand governments managed geothermal electricity generation from the 1950s to the1980s. Their management of the TVZ resource rarely took into account Māori values.

As the Waitangi Tribunal has found, government management caused physical impacts at Ohaaki and Wairākei-Tauhara at Taupō. In both cases, extraction caused subsidence throughout the Ohaaki and Wairākei-Tauhara area. At other sites, such as Orākei-Korakō, major features were flooded through hydroelectric development and associated dams on the Waikato River. At Kawerau, Lake Rotoitipaku was used as a dumping pond. The result was that important cultural sites were lost and, with the loss, Māori knowledge and their ability to sustainably utilize the geothermal resources on their lands was undermined. However, that situation has changed since the restructuring of the electricity generation and supply industry and the movement of development away from the government to the private sector. The need for resource consents under the RMA has also pushed the electricity sector and Māori to work together to avoid lengthy litigation before the Environment Court.

As a result, Māori are now significant players in the industry.[21] By 2015, the landowners of the fields at Tauhara North, including Rotokawa, Ngātamariki and Mōkai, together with Mercury Energy, were among the world's largest geothermal operators, calculated according to installed capacity.[22]

In January 2017, I interviewed Nick Clarke, the general manager geothermal for Mercury, and Bruce Waters, the company's general counsel.[23] Clarke has been at the company since 2015, and Waters began with the company when it was set up in 1999 as Mighty River Power.[24] The company rebranded itself as Mercury and effectively retired the Mighty River Power name in 2016. Mercury has the largest shareholding base of any company in New Zealand, and it services about 100,000 homes and approximately

More than 17 percent of New Zealand's total electricity supply comes from geothermal generation. Most geothermal resources are subject to ownership claims by local Māori, including those used by the Ohaaki Power Station near Taupō, North Island, New Zealand. *Photograph by Alan Gibson, New Zealand Herald*

350,000 customers. According to its marketing, it seeks to "inspire New Zealanders to enjoy energy in more wonderful ways."[25]

In terms of its management of geothermal stations, Waters explained the mutual dependence the company shares with their joint venture partners in geothermal development. Thus, building relationships was and is an essential ingredient in the successful development of the geothermal power stations operated by the company. This approach was inspired and promoted by Doug Heffernan, the founding CEO. His leadership was augmented by board members such as Tania Simpson, the longest serving director (2001–15), a tribal member of the Ngāti Maniapoto tribe and Tainui Waka near Hamilton.

Along with its relationships, the foundations of the company are outlined in its annual reports, such as one it produced in 2016.[26] In that report, the company described its business model foundations as "(1) Well-being (2) Kaitiakitanga (Guardianship or Stewardship) (3) Commercial." Mercury Energy declared that "through its licence to operate, relationships with communities and its commercial partners, it will sustainably harness resources to generate low-cost renewable electricity." In annual reports since 2016, and while other foundation stones of the business have changed or been rewritten, the value of *kaitiakitanga* has remained a part of the company's business model.

When asked whether the company can exercise *kaitiakitanga* without its current Māori partners, both men agreed that the practice of *kaitiakitanga* and the values that underpin it, including sustainably managing the resource for future generations, are consistent with the company's focus. Waters added that it was not possible to exercise *kaitiakitanga* without their joint venture Māori partners, and that *kaitiakitanga* aligns with the notion of "sustainably harnessing" the resource.

In describing the resources of a geothermal field, Clarke echoed the earlier observations of Hāmuera Mitchell Sr. about the nature of the resource: "The reservoirs are living—you know, almost in the literal sense. They change and they come and they go and we try and understand it so we spend a lot of time picking where to drill."

Both Mitchell and Clarke are concerned with the nature of the resource and how it functions. For the company, it is just as important as it is to Māori to ascertain the size of the reservoir and to protect it to ensure its sustainability. In doing so, three things are important to the development of electricity: heat, water, and pressure. If any of those factors is missing, there can be no development. Māori also need all three elements to continue their customary relationship with the resource. There are synergies that neatly form the basis for joint ventures where electricity generation companies take Māori values seriously. In this case the concept of *kaitiakitanga* demanded by Campbell and her people, and the management plans instituted by the joint venture partners, have improved their relationship with each other and the resource. It has also ensured the continued sustainable development of their joint geothermal operations.

Mercury Energy's power stations have been built with reinjection technology, meaning they can reinject geothermal fluid following processing. Reinjection has been used as a means of sustaining the geothermal fields upon which its power stations are located. As with the Ohaaki Power Station, location of sites for reinjection are important. Mercury has a team of approximately 100 geoscientists, geochemists, and engineers constantly monitoring the resource. They undertake reservoir modeling to try and understand what is happening underground and to predict how the resource will respond to development over time. Data on the state of the fields is extrapolated from the wells used for this modeling. In this manner, the company's understanding of the reservoirs is consistently improving.

As the interview with Campbell demonstrates, the relationship with Mercury Energy has been important. Instrumental to the nature of that relationship has been understanding the Māori value of preserving the resource for future generations. In other words, sustainability seriously underscores the strength of their joint ventures. Nick Clarke, for example, noted that prior to the building of the Ngātamariki Power Station, there was a concern around whether there would be any link between the development of the Ngātamariki reservoir and the remaining surface features at Orākei-Korakō. Due to the relationship of the company with Ngāti Tahu and Ngāti Whaoa, and its interest in sustaining the resource, the company chose to use state of the art reinjection processes, which results in 100 percent of the fluids being reinjected back into the reservoir, rather than losing pressure through evaporation and potentially impacting those remaining features. The company also has monitoring wells between its operation at Ngātamariki Power Station and Orākei-Korakō, and it must report annually to a peer review panel as a further precautionary measure.[27]

Māori concerns regarding Orākei-Korakō, Ohaaki, and other sites that form the geothermal resource of the TVZ relate to their cultural understanding that the TVZ is part of an indivisible metaphysical system, one that is personified by the various elemental gods recognized in Māori cosmology.

The gods and the early ancestors have become features of the Māori management narrative. It is the basis upon which they exercise rights to land and natural resources including geothermal resources, and with those rights come the responsibility to care for the resource as guardians for the future generations yet to come. The desire to preserve the geothermal fields for future generations has been a driving factor in underpinning Māori participation in geothermal management procedures and, as a result, reinjection of geothermal fluid in the right places within a field has become a central feature of geothermal power generation. Reinjection is now accepted in ventures involving geothermal power development,[28] and those who harness the resources of the TVZ should now understand Māori traditions and values.

The traditions and oral history as recorded by the Waitangi Tribunal, and the oral history of contemporary actors such as Aroha Campbell and the comments of the Mercury leadership team, provide a platform for a better understanding between two cultures about how to manage the geothermal features of the TVZ, and it has, or will hopefully, inform better management practices of developers. In this small way, the recording of Aroha Campbell and the interviews undertaken with the Mercury leadership team members demonstrates how indigenous peoples' knowledge of the natural environment benefits all of us and the future generations to come. The previous lack of recognition by successive governments, the marginalization of Aroha Campbell and her community, and the desecration of their important sites at Ohaaki were all caused by unsustainable practices in the generation of electricity. Such marginalization was overcome only when good leadership from Māori and the companies concerned demonstrated a commitment to building a relationship with each other and conducting their mutual business based on sustainable practices. This collaborative approach has resulted in the integration of Māori tradition, oral history, and values, such that now the geothermal resource of the TVZ is subject to an improved management ethos.

The Aotearoa/New Zealand experience in the management of geothermal features reminds all countries to listen to the voices of their indigenous peoples and ethnic minority populations that live on the land, by their rivers, mountains, and seas. To attain good environmental outcomes, their values, combined with the expertise and goodwill of developers, may hold the key to the sustainable management of natural resources and result in all of us accepting the responsibility of being guardians of the natural world for future generations.

Notes

1. Waitangi Tribunal, *He Maunga Rongo – Report on the Central North Island Claims* (Wai 1200, Legislation Direct, 2008).
2. The Waitangi Tribunal described the nature of the subterranean resource of the TVZ as follows:

    "Within the TVZ are some 17 major geothermal fields where surface water penetrates deep into the earth, is heated by encountering magma which has intruded into the fractured crust of the TVZ, and emerges as boiling water or steam. The pressures within the TVZ have increased the temperature of rocks below the surface, and fractures within the rocks allow surface water to penetrate deep into the earth. Each of these geothermal fields has its own unique combination of surface features, including hot springs, mud pools, geysers, fumaroles, and sinter deposits."

3. I note that the Ngāti Whaoa tribe and Ngāti Tūwharetoa also have interests in many of the geothermal fields discussed in this chapter.
4. The Resource Management Act 1991 (RMA) is the primary statute in New Zealand governing the management of natural resource use and development in New Zealand. It allocates responsibility for the management of these resources to local authorities with appeal procedures to the Environment Court. Developers require a range of resource consents for major developments such as geothermal power stations. Part II of the RMA requires decision makers to recognize and provide for the relationship of Māori and their culture and traditions with their ancestral lands, water, sites, *waahi tapu* (sacred sites), and other *taonga* (treasures). They must also have particular regard to *kaitiakitanga* (guardianship) and take into account the principles of the Treaty of Waitangi.
5. Effectively the land is sinking, resulting in inundation from the river, although it has slowed down. The local Māori people will have to negotiate the removal of the *marae* if the subsidence is recorded below the agreed level.
6. It seems that the problem at Ohaaki for the period 1988–1998 was the method of and placement of reinjection. See Sanggoo Lee and Lew Bacon, "Operational History of The Ohaaki Geothermal Field, New Zealand," *Proceedings World Geothermal Congress 2000* Kyushu - Tohoku, Japan (May 28–June 10, 2000), 3211–16.
7. Campbell refers to Rotokawa Stage I. It was bought by Mighty River Power and operated as part of the 50:50 joint venture between Mercury Energy and Tauhara North 2 Trust, known as the Rotokawa joint venture.
8. The NAP Power Station (Nga Awa Purua) is a joint venture with Tauhara North No.2 Trust. The trust owns 35% and Mighty River Power (now Mercury Energy) owns 65%.
9. More details on the Rotokawa II/Nga Awa Purua geothermal power station features, including its production wells and steam turbine, can be found here: https://www.renewable-technology.com/projects/geothermal-power-plant.
10. The $466m plant is fully owned by Mercury. The fuel rights of the plant are jointly owned by Mercury and Tauhara North No.2 Trust.

11. See John Te H. Grace, *Tuwharetoaū: The History of the Māori People of the Taupō District* (Dunedin: A. H. & A. W. Reed, 1959), 407.
12. As recorded in Waitangi Tribunal, *He Maunga Rongo – Report on the Central North Island Claims* (Wai 1200, Legislation Direct, 2008) Vol. 4, chapter 20, 1494.
13. Ibid., 1488.
14. Ibid., 1536–38.
15. Te Awa Tupua (Whanganui River Claims Settlement) Act 2017, Part 2 & Subpart 2.
16. More information can be found here: www.waikatoregion.govt.nz/Community/Your-community/iwi/A-Maori-Perspective-Te-Ao-Maori/Maori-and-geothermal-resources/ and https://www.boprc.govt.nz/your-council/plans-and-policies/plans/regional-plans/rotorua-geothermal-regional-plan.
17. The Māori version of the Treaty of Waitangi was signed in 1840 between the British Crown and over 500 Māori chiefs representing all the tribes of New Zealand. Principles of the Treaty of Waitangi have been incorporated into several statutes of the New Zealand Parliament including the Treaty of Waitangi Act 1975, which set up the Waitangi Tribunal, and it is also referenced in section 8 of the Resource Management Act 1991.
18. Waitangi Tribunal, *He Maunga Rongo – Report on the Central North Island Claims* (Wai 1200, Legislation Direct, 2008) Vol. 4, chapter 20, 1484–85.
19. More information on the history of use and management can be found here: https://www.boprc.govt.nz/environment/geothermal-resource/kawerau-geothermal-system/.
20. Ibid.
21. Since Wairakei, there have been several other geothermal stations constructed—these include Ohaaki, Rotokawa, Poihipi, Mōkai, Kawerau, Ngātamariki, Tauhara 1 (Te Huka), GDL (at Kawerau), Ngā Awa Purua, Te Mihi, and Topp 1 (at Kawerau). Six of these are operated by Mercury, and five are owned or part-owned by the same company.
22. Tania Simpson Speech, Energy Matters Conference (Auckland University, NZ, June 18, 2015) https://www.youtube.com/watch?v=QtKEglz8eQc
23. January 17, 2017.
24. Mighty River Power was a product of the state sector reforms in the 1990s, when the political decision was made to break up Electricity Company of New Zealand (ECNZ) into four parts. One of these, Contact Energy, split off in 1996. In 1999, the three state-owned enterprises, Meridian, Genesis, and Mighty River Power, were established. Mighty River Power then acquired the ECNZ's assets. The company has been managed by a board with a shareholding minister of the New Zealand government. Mighty River Power has rebranded its entire operation as Mercury Energy. When the transfer of assets from ECNZ took place, the company had part ownership of the Mōkai Geothermal Power Station with the Tuaropaki Māori Land Trust, and it operated the small station at Rotokawa. Those were the only two operations at that point. Since then, generation capacities at Mōkai and Rotokawa were increased with the construction of Mokai II and Ngā Awa Puru power stations respectively. New geothermal stations were also constructed at Kawerau.

25. Mercury Energy – 2016 Annual Report, 4.
26. Ibid., 18–19.
27. In terms of the future activity of Mercury Energy, currently the supply of geothermal power is sufficient to meet the country's needs and, unless there is increased demand for electricity in New Zealand, there will not be any need to enter into any further joint venture arrangements with any new Māori landowners for some time. However, as the demands for electric cars and other new applications increase, that situation may change.
28. See the Tauhara North 2 website at www.tauharano2.co.nz.

9

When Little Fish Encounter a Big Dam

Environmental Conflict on the Upper Yangtze

Dai Qing and Kang Xue

Oral history by Kang Xue, retired Beijing People's Radio journalist, with Zhang Boju, secretary-general of Chinese nongovernmental environmental protection organization, Friends of Nature, Chongying, China, June 2016

Founded in 1994, Friends of Nature was the first green NGO created in China. The group has served as an incubator for an explosion of environmental organizations. Friends of Nature has worked to inform Chinese citizens on critical environmental issues through a host of educational programs, policy reform initiatives, and community advocacy. The Friends of Nature Greenbook, *published annually, provides the definitive record of the ongoing top environmental issues in China. Zhang Boju is the current CEO of Friends of Nature and leads a team of more than 20,000 members working to preserve fragile Chinese ecosystems and promote the development of more green citizens in the rapidly developing country.*

BOJU: In 2009, I was working in the research department of Friends of Nature, a Chinese environmental protection NGO. In my role, I was responsible for responding to cases of environmental destruction. One day in May, the then secretary-general, Li Bo, gave me an assignment: to prevent the modification of the boundaries of the Upper Yangtze Rare and Endemic Fish National Nature Reserve that was required before regulatory approval could be given for the construction of the Xiaonanhai hydroelectric plant.

After quickly gaining an understanding of the environmental context of the Xiaonanhai hydroelectric project, my reaction was that it would be

Dai Qing and Kang Xue, *When Little Fish Encounter a Big Dam* In: *Oral History and the Environment.* Edited by Stephen M. Sloan and Mark Cave, Oxford University Press. © Oxford University Press 2022.
DOI: 10.1093/oso/9780190684969.003.0010

absurd to modify the boundaries of the nature reserve. How could this possibly be on the table?

Xiaonanhai is the name given to a section of the Yangtze River in the Chongqing area, in the upper reaches of the river and just above the 600-kilometer length of the Three Gorges Dam reservoir. This is the richest stretch of river in China for the country's rare and endemic fish species. Since the construction of hydroelectric plants along the Yangtze began in the 1980s, thirty-five enormous dams have been built—can you imagine? The completion of Xiaonanhai would take that number to thirty-eight.[1] It would also result in the obstruction of the migration paths of many species of fish. In order to spawn, these fish make extraordinarily long journeys.[2]

The Xiaonanhai project is directly related to the Three Gorges Dam, which began construction in 1994. In order to reduce the negative impact of the Three Gorges' construction work on the river's fish population, a protected nature reserve for rare and endemic fish species was established in the Upper Yangtze. In April 2000, this nature reserve was upgraded to the national level, becoming the Upper Yangtze Rare and Endemic Fish National Nature Reserve. The proposed site of the Xiaonanhai dam is an important part of this protected zone. According to experts, approximately fifteen billion young fish and spawn pass through this section of the river every year.

I described the situation as "absurd" because the boundaries of this national-level nature reserve had already been modified once in order to accommodate other hydroelectric plants in the Upper Yangtze—the reserve was described as having had its "head" pinched off. At the time, it was reported that this was on the order of a "high-up authority," meaning that after that modification, there would be no further changes to territory of the reserve. But in January that year, Chongqing Municipal Government announced yet another controversial development: a plan to construct three large hydroelectric dams at Xiaonanhai, Zhuyangxi, and Shipeng, effectively cutting off the tail of the nature reserve. It looked likely that their plan would be executed. In February and May, the Ministry of Agriculture supported the Chongqing Municipal Government's proposal to modify the nature reserve before handing the issue off to the Ministry of Environmental Protection.

The first action opposing this plan had already been taken on May 6, before I accepted the assignment on behalf of Friends of Nature. Eight experts on Yangtze River wildlife and the ecology of southwest China, including Peking University Protected Species Professor Lu Zhi, published an open letter in which they called for "the prevention of excessive hydroelectric development and the protection of rare and endemic species of fish."

The most earnest efforts to prevent the construction of the planned dam were taken between August and the end of the year. We came up with a strategy to inform the public and encourage citizens to voice their concerns about the plan. By engaging public opinion, we hoped to dissuade the National Nature Reserve Review Committee from approving the modification of the Upper Yangtze Rare and Endemic Fish National Nature Reserve and prevent Chongqing Municipal Government from obtaining a construction permit for a hydroelectric dam at Xiaonanhai. Friends of Nature gathered together other environmental protection organizations and we jointly issued an open letter to the Chongqing Municipal Government.

On August 31, I learned that the Ministry of Agriculture had applied to the Ministry of Environmental Protection to modify the boundaries of the nature reserve. In response, we began a series of actions: we wrote an open letter to the experts of the National Nature Reserve Review Committee, "requesting that each expert on the committee holds fast to their academic integrity and casts a vote that will stand the test of history." We did not have a list of the experts' names, so we requested that the Ministry of Environmental Protection make the list public. In our capacity as an environmental protection organization, we asked to audit their meetings. We also requested that they publish the minutes of all their meetings, as well as the evaluations and opinions of each expert, and asked that the experts accept our oversight.

We also became more assertive in our media campaign. I remember making daily phone calls to journalists to explain the significance of this matter, inform them of each new development, and promote the meeting of the National Nature Reserve Review Committee as a hot news item. We revealed the time, location, and agenda of the meeting, to ensure that it would not become a "closed meeting," unknown to the public.

My communications, interactions, and cooperation with the Ministry of Environmental Protection left a particularly deep impression on me, especially those with the Nature Reserve Administrative Office of the Ecology Department. At that time, as a young person only a few years out of college, I didn't know many government officials, and so all I could do was telephone their office. They had a very positive attitude, which surprised me. Even when it came to our request to audit their meetings—which actually had no legal basis—they said: "This request is unprecedented, but we promise to make all information on Xiaonanhai available to you, going forward." I particularly remember that on November 10, an official called me to say: "The topic of Xiaonanhai was not added to this meeting's agenda in the end." This meant that the decision had been postponed—happy news!

Only afterwards did I come to understand that the Chongqing Municipal Government could not give up on its plan to build the three dams at this point. In the following year, meetings took place without our knowledge. At the Ministry of Environmental Protection's Expert Review Meeting at the end of 2010, without the knowledge of the environmental protection organizations, the proposal to modify the nature reserve was unanimously passed.

I cannot describe how angry I was at that time. Because Xiaonanhai would no longer be within the nature reserve, we lost the legal basis for our opposition to the construction of a dam in that area. Over the phone, I questioned the official from the Ministry of Environmental Protection: "Did you disclose all the information, like you promised? Was the proposal passed unanimously? Including by the experts who had opposed it all along?"

After hearing the helplessness in his voice, I stopped questioning him. I also finally grasped that this issue was not that simple and understood the power of Chongqing Municipal Government. At the end of 2007, Bo Xilai assumed his post as member of the Central Politburo and simultaneously as Communist Party Secretary of Chongqing. The Xiaonanhai hydroelectric plant was one of his most valued engineering projects. With a projected investment of 32 billion RMB, it was the largest engineering project in the history of Chongqing and was certain to raise GDP as soon as construction began. For this reason, neither the Ministry of Agriculture nor the Ministry of Environmental Protection nor the experts and environmental groups could get in the way. For the first time, I felt that we resembled the young fish of the Upper Yangtze—insignificant, being pushed out by an enormous dam. We felt suffocated.

In the moments of deepest despair, a senior official from the Ministry of Environmental Protection gave me encouragement and taught me about the method of "stalling for time." Several times over the phone, he said: "Little Zhang, don't be discouraged, just stall for time and keep stalling until something changes. We are also stalling for time." I understood that as an official within the system, all this senior official could do was cause procedural delays and postpone the beginning of construction of the Xiaonanhai project. As a non-governmental environmental organization, as long as construction hadn't yet started, we would spare no effort to prevent it.

In February 2011, during the spring festival, I flew directly to Chongqing, wanting to see Xiaonanhai with my own eyes. As soon as I left the airport, I received a call from an unknown phone number: "Mr. Zhang, welcome to Chongqing. If you have time, please do come and see us." I was startled. It was clear that I was being watched! That happened on the fourth day of the Lunar New Year, while the whole country was still on vacation.

I didn't know what was going to happen, but I was determined to proceed according to my plan. However, I was careful to stay in a small, eighteen RMB-per-night hotel that didn't require guests to show their ID card. Early next morning, under orange street lamps, I headed to the ferry crossing and finally saw the Yangtze River and Xiaonanhai for real.

I took the first ferry across the river and got off at Zhongba Island, the proposed site of the Xiaonanhai hydroelectric plant. The third largest island in the Yangtze mainstream, some say it resembles a fish—the vast dam of the Xiaonanhai hydroelectric plant would straddle this fish's head.

I didn't dare remain in Chongqing. I rushed around in a hurry all day. I traveled by all kinds of transportation—coach, minibus, motorcycle. In the evening, in a small restaurant in Luzhou in Sichuan Province, I drank a bowl of liquor, alone, with a kind of disquiet in my heart. Until now, I haven't openly spoken about the profound unease I felt in Chongqing.

I made the best of my time. In Hejiang County, Sichuan Province—the core of the nature reserve—I visited several fish restaurants and asked the proprietors: "Have there been many fish recently? Any good fish?[3] Any 'big red' fish?"[4]

A PROPRIETOR REPLIED: "Not many, but some."

"Do you know that it's a protected species?"

The proprietor began to laugh. "Hey, customers want it! If we can get it, it's good for business!" he said.

I also asked if people still fished with electricity.[5] He replied: "Very few."

Later, a member of the Chongqing branch of Friends of Nature, Mr. Pu Kaiyun, told me that these fish restaurants would not sell a protected species of fish to an unknown, out-of-town customer. Regular, local customers would ask them: "Boss, do you have any better quality fish? Can I take a look?" The proprietor would take them out back to look. Apparently the rarest and most expensive species would fetch more than 1000 RMB for one *jin* of fish meat.[6]

In any case, this demonstrated that the rare fish species of the Yangtze River were still present in Xiaonanhai. This was the most significant finding of my trip. After the extinction of the Chinese sturgeon, the major protected species of fish were the Dabry's sturgeon, the Chinese paddlefish, and the Chinese high fin banded shark. These were the same fish referred to by those in favor of the dam—they talked about groups like ours "opposing a big dam with three single fish." They consistently challenged attempts to protect the species, saying: "The protected fish species of the Yangtze have already gone! What is left to protect?"

As a non-governmental organization, Friends of Nature has always suffered from a lack of channels through which to influence policy. In our most anxiety-filled days when the nature reserve was being modified, the young staff of Friends of Nature were truly worried! We decided to submit a proposal entitled "The Losses of the Xiaonanhai Hydroelectric Plant Outweigh the Gains" to the 2012 "two sessions,"[7] the highest organs of state power.

The "two sessions" were set to begin in March 2012. During the whole winter leading up to them, my colleague Dou Lili had, to our surprise, contacted more than a hundred representatives and committee members who might be able to help Friends of Nature submit our proposal. In the end, Professor Sheng from a university in Beijing gave us a clear reply: "I study ecology and understand how destructive this project will be." He submitted the proposal. Who could have predicted that the following day, he would withdraw it, claiming that "the facts are not fully established and the evidence is insufficient"? Journalists made much of this—after all, proposals are rarely withdrawn, especially those relating to Chongqing. At that time, Chongqing was in the national spotlight: the Deputy Mayor and Police Chief Wang Lijun had recently emerged from the American embassy in Chengdu, and Bo Xilai's leadership was coming under scrutiny. The Chongqing authorities may have wanted to avoid further attention, which our proposal might have generated. Afterwards, I heard a rumor that on the evening when our proposal was submitted, two members of Chongqing's delegation paid a visit to Professor Sheng to drink tea and have a chat.[8] I know that at that time Professor Sheng was also the Party Secretary of his university. I am still grateful to him today for the help he gave us, impelled by his professional conscience.

On March 21, after the NPC and the CPPCC sessions, I took my second trip to Xiaonanhai and once again alighted at Zhongba Island. It was spring, and the island was teeming with wildlife. But a large piece of land had been cleared and flattened, and they were erecting a stage with a metal frame. I asked someone and discovered that the site was being prepared for the ceremony marking the beginning of construction on the Xiaonanhai hydroelectric project. On March 29, 2012 the foundation stone-laying ceremony for the Xiaonanhai hydroelectric project would take place here. The several experts and journalists that were traveling with me began to discuss this development—things were moving so quickly! The project's environmental impact assessment had not even been started.

Though the foundation stone-laying ceremony was held according to schedule, Bo Xilai (who would not serve another term as Party Secretary of Chongqing) did not attend, and construction of the Xiaonanhai hydroelectric plant did not begin. It kept on getting postponed.

April 8, 2015, is a day that I will never forget. On the morning of that day, I had a meeting scheduled with colleagues. On the way to the meeting, I received a WeChat message from a journalist, and in a fluster I saw that it was a document issued by the Ministry of Environmental Protection on March 30, titled, "An Official Reply Regarding the Environmental Impact Assessment of the Jinsha River Wudongde Hydroelectric Plant." I didn't pay much attention to it and continued on my way, but the messages kept coming; the journalist wanted a comment from me. At that point, I discovered the huge news hidden in the document:

> Within the borders of the nature reserve from the Xiangjiaba Dam to the end of the Three Gorges Reservoir in the mainstream of the Yangtze River and its tributaries, the Minjiang River and the Chishui River, further planning and construction of the Xiaonanhai Hydroelectric Plant, Zhuyangxi Hydroelectric Plant, and Shipeng Hydroelectric Plant and any other dams or water-related engineering projects are prohibited.

This was a calm and cool attack. Out of the blue, the Ministry of Environmental Protection chose this strategy, hiding their rejection of the Xiaonanhai hydroelectric project within the assessment report of a different hydroelectric project! After receiving several interviews over the phone, I began to cry. I wept audibly on the subway.

In a speech on January 5, 2016, China's highest leader, Xi Jinping, described the Yangtze River as an important ecological treasure house and celebrated its unique ecosystem. He spoke of the necessity, now and in the long-term future, of restoring the Yangtze's ecology and environment, and called for an emphasis on protection and a rejection of over-development. Is it a coincidence that he gave this speech in Chongqing?

Countless times I have thought back to 2009 and the beginning of our successful campaign to stop the modification of the nature reserve at Xiaonanhai. At the time, perhaps the Chongqing authorities wanted to avoid being in the limelight on account of a controversial new hydroelectric project and thought there would be no harm in delaying. But because that delay coincided with the period leading up to the fall of Bo Xilai, construction of the Xiaonanhai dam was never started. Without that delay, Xiaonanhai hydroelectric plant would probably have already begun producing electricity—and that would have been a tragedy for those little fish.

Commentary by Dai Qing

Reading this oral history, one is led to reflect on the early twenty-first century emergence of a certain set of attitudes among powerful and prosperous Chinese people, both on and off the global political stage: extravagance, triumphalism, arrogance, and chutzpah. In the nineteenth century, and even up to the final years of the twentieth century, this was not a common sight. When one asks why, the answer given by observers—whether in admiration or antipathy—is always the same: "We are witnessing the rise of a great power!" The term *rise* invokes the rapid growth of the GDP and "the second superpower," which is ever advancing and able to mobilize troops that inflict heavy casualties . . . all true.

But the problem is this: are these things dropped down from heaven, simply bestowed upon us? Or, to put it another way, does this incomparably arrogant "rise" come at a cost? If so, how do we pay for it? And who, exactly, pays for it? In the twenty-first century, the international order is no longer constituted by the British Empire forging pathways and establishing colonies in other lands with gunboats. The "rise" of China has relied on communist autocracy and the extortion and plundering—of capital, property, and rights—that occurs within its system. The fact that China is a country with vast territory and an enormous population is always cited as justification. But the flashy novelty and prideful self-indulgence flaunted by the rich and powerful are bought with the most vulnerable and lowly elements of society, who have no voice: the migrant workers who are unaware of their basic rights, and the silent earth—forests, mineral repositories, soil, rivers.

However, the ruling party knows that educated people with international outlooks and sensibilities refuse to be silent. In terms of what they embrace and reject, their knowledge, their courage, and their sense of responsibility, Chinese environmental NGOs resemble environmental groups around the world. The difference between them lies in the political rights of environmental activists in China: Can they gain access to facts? Can they issue statements? Can they assemble? Can they plan and organize? Can they receive donations? Can they enter the rooms in which the government policy is formulated and state their views?

It cannot be said that the Chinese environmental NGO Friends of Nature—which is still unable to register officially as a social association—has been utterly stripped of those rights. But they must "self-censor" at all times.

Thus fettered, current secretary-general Zhang Boju relinquished the pleasures and material comforts pursued by most young people of his generation. Instead, he entered Xiaonanhai, the site of yet another disastrous mega-dam project. In his interview he remarked how unimaginable was the prospect of building another structure on a river that had already been dammed thirty-five times upstream of the Three Gorges Dam. Also, the need for a rare fish reserve has been debated at length from all sides of the issue and has been settled. Boju believed that erudite professors and experts would not trample on their own knowledge and conscience in the face of power and profit. If that was not enough, it seemed politicians would hold high the importance of fish, fishermen, and the forests and rivers of the nature reserve to block the project. How could he have known that most of the Upper Yangtze dams were constructed primarily to prevent the buildup of silt in the Three Gorges? This occurred only after experiments revealed that the Three Gorges Dam would suffer significant silt deposits, a design flaw that officials had a big incentive to cover up by constructing the other dams. For governmental planners, the fish species about to become extinct mattered little in comparison to the profit and promotions officials stood to gain by constructing a dam and boosting GDP.[9] Unfortunately, the rule of law, procedure, and any sympathy for or desire to placate environmentalists felt by a few officials counted nothing next to the "closed meetings" at which high-level powers divide the spoils.

Everyone knew that the stakes of the Xiaonanhai project were catastrophically high. Ultimately, the unhappy political fate of its champion transformed the project from a policy directive that was futile to resist to a shelved scheme treated with indifference.

In this oral history, the repressed anger and helplessness of the environmental activists are tangible. The gap between what they want to do and what they are permitted to do is so large that it is difficult to comprehend. Confronting reality, the narrator and his group acted with restraint: not questioning the institutional impudence they encountered, not probing the deals made between the powerful and the wealthy, not insisting on their right to public oversight. They only made their talking points milder and milder. In the end, the fate of Xiaonanhai was decided by higher powers, and the outcome seems to have had absolutely nothing to do with this group's efforts. Like their fellow environmental groups around the world, Friends of Nature conduct investigations and do research, analyze information, and seek opportunities to provide input to policymakers. In today's China, however, they are merely cards in the hands of politicians, ready to be played. The fate of China's most precious resources and environments, for better or worse, are determined by the consequences of political power struggles at the highest levels of government.

The Three Gorges Dam project should be taken as a warning. In the long river of history, the fall of a dynasty due to a power struggle represents nothing more than a brief clamor. But what about the rivers, forests, and ever-vanishing species?

The narrator gave his youth to an arduous—and perhaps silent and traceless—act of resistance. In a country where the political rights of citizens are worth less than a piece of wastepaper, I will not dare to give encouragement to these groups and the young people who comprise them.

The new law[10] has defined foreign NGOs working in China as foreign powers with an insatiable desire to subjugate us.[11] China's environmental NGOs can still pick up trash and go birdwatching. However, when it comes to activism that directly touches on high-level political and financial interests, encouraging young people to muster up the courage and determination to act is tantamount to asking them to charge into gunfire. I do not think I have the right to do so.

But those who have chosen to walk that dark and difficult path deserve understanding and support.

We know that the largest sacrifices made in the name of China's rise are its natural resources, its environment, and its "silent majority." They are silent, first because they lack opportunities to speak. This majority also lack the education and social skills necessary to recount their experiences. While narrators of oral histories raise deeper and deeper questions, the silent majority does not recognize or even understand that their own "trivial" experiences—their own deeply buried emotions and desires—constitute China's most fundamental elements, its DNA.

What about the narrators of oral histories? When one governing party with a single authority figure at its core monopolizes power, official state media is simply a mouthpiece, just as it has been for since 1949; the military police can use guns to deal with a China that tries to speak or even think, just as they could since 1949. In today's China, these narrators show their courage and wisdom by pursuing life and truth in the narrow space that they occupy. They are willing to struggle to inform the rest of the world that this China is in fact an "iron house" that looks new and shiny only from the outside. They also speak to their fellow compatriots living in the "iron house" of what fresh air is . . . never seeking recompense (just compare the price of an environmental oral history and a trashy pornographic novel today). What they do not realize is the great value and critical importance of their stories in the context of China's modernizing transformation.

Fighters, in the midst of frustration and exhaustion, keep fighting!

Notes

1. Two other hydroelectric dams at Shipeng and Zhuyangxi, located immediately upstream from Xiaonanhai, were also planned for construction as part of the same project.
2. In order to complete the reproductive process and for their spawn to develop into fry, the fish species that reproduce in the nature reserve must travel between 200 and 500 kilometers in the right conditions—water of a certain temperature and with a particular speed of current. The spawn of these fish pass through Xiaonanhai and are drawn downstream into the Three Gorges Dam area by the current. After developing into fry,

they must return upstream via Xiaonanhai. The Xiaonanhai dam would have been a barrier to the upstream and downstream migration of fish, and therefore an impediment to their reproduction.

3. For local fishermen, "good fish" usually means relatively rare and difficult to catch fish. These are more precious and can be sold at a higher price. These may be, but are not always, protected species.
4. "Big red fish" is the local nickname for the Chinese high-fin banded shark.
5. Fishing with electricity—also known as electrofishing—is a method of fishing in which an electric current is applied to the water in order to stun fish, making them easier to scoop up in large quantities. Once popular in China, electrofishing has been linked to the depletion of fish stocks in the country's rivers and was outlawed in 2012.
6. Just over a pound. One *jin* = 1.10 pounds.
7. The "two sessions" are the National People's Congress and the Chinese People's Political Consultative Conference, the two main deliberative bodies of China, which meet annually to make national-level political decisions.
8. In China, "to be invited for tea"—or to "drink tea and chat"—has become a well-known euphemism for being questioned by the authorities. The implication here is that Professor Sheng was asked to withdraw the proposal by the Chongqing authorities.
9. The "profit" that Dai Qing refers to is that which goes to the "benefit groups" who promote such large-scale projects in the form of graft, sweetheart contracts, and monopoly pricing, and does not refer to the overall profitability of the project. In fact, if all costs are internalized, large dam projects generally lose money and do not generate profits. Another way for officials to "profit" from large dam projects is to hive off the revenue-generating, power-producing facilities into separate companies in which they are shareholders or managers, or both. Meanwhile, the dam infrastructure that requires ongoing costly expenditures—environmental stabilization and remediation, resettlement, etc.—remains in public hands and continues to drain the public purse. Authorities also "boost" GDP figures with public expenditures that are undisciplined by market forces or public oversight, making GDP data an unreliable measure of the creation of real wealth.
10. In April 2016, the Chinese government passed the "Overseas NGO Management Law" to regulate foreign nongovernmental organizations operating in the country. The law came into effect on January 1, 2017, and stipulates that foreign organizations (including advocacy groups, business chambers, and foundations) must register with the local public security bureau and link up with a pre-approved Chinese partner or sponsor organization.
11. In her interpretation of the new law, the author wryly quotes Mao Zedong's description of foreign imperial powers as having "an insatiable desire to subjugate China."

10

The Free Play of Natural Forces

Wild Methods of Oral History in Documenting Wilderness

Debbie Lee

Oral history by Debbie Lee with Wilderness Ranger Art Seamans, Lewiston, Idaho, April 28, 2013

From 2010 to 2014, I directed the Selway-Bitterroot Wilderness History Project, whose mission was to chronicle the human history of one of America's largest wilderness areas. Funded by the National Endowment for the Humanities, the project was a joint effort between Washington State University, the University of Idaho, and the US Forest Service. My collaborator, Dennis Baird, was charged with collecting more than sixteen cubic feet of documents. I handled the "live" history, interviewing mule packers, backpackers, river rangers, river outfitters, Forest Service employees, backcountry pilots, a Native American tour guide, artists, educators, scientists, archivists, volunteers, and caretakers of inholdings (three privately held parcels within the wilderness boundary). We wanted to preserve the human history of the wilderness with the same passion that people had used to preserve the land itself. It turned out that my interviewees were equally enthusiastic about that goal. One of them, Art Seamans, worked as a wilderness ranger in the Selway-Bitterroot from 1975 to 1980 at the remote Moose Creek Ranger Station.[1] *He was seventy-seven when I interviewed him.*

Funding for the research and writing of this chapter was provided by the National Endowment for the Humanities Collaborative Research Grant for the Selway-Bitterroot Wilderness History Project. Gratitude to my co-PI for the project, Dennis Baird, to our students and assistants Erin Jepsen, Anna Conditt, Jane Holman, James Trout, and Ben Bunting, as well as to our partners at the US Forest Service Nez Perce Clearwater National Forest, particularly Cindy L. Schacher.

Debbie Lee, *The Free Play of Natural Forces* In: *Oral History and the Environment*. Edited by Stephen M. Sloan and Mark Cave, Oxford University Press. © Oxford University Press 2022. DOI: 10.1093/oso/9780190684969.003.0011

LEE: Can you tell me how you first heard about the Selway-Bitterroot and what got you interested in that particular landscape?

SEAMANS: Well, I was working with blister rust control [applying fungicide to blister rust, a disease affecting white pine throughout the US] when I started with the Forest Service. I started 1959. Blister rust crews were the IR fire crews [inter-regional fire suppression crews] at the time, really. In 1961, we had a very severe fire season and I was on a crew that was sent up the Boulder Creek area to the Surprise Creek fire. While we were fighting the fire, I was out scouting line with another individual and we got to the top of a ridge and we were looking down into Moose Creek from that ridge top, and I saw there was a ranger station down there and said, "Boy! I would like to work down there someday. That looks like my kind of place!"

LEE: So, when you say "my kind of place," what is your kind of place?

SEAMANS: Well, it was wild. It didn't have any roads, it was very isolated, and I love the backcountry. When I worked with blister rust I spent one summer doing surveys pretty much all over the Clearwater National Forest [in Idaho] and some of the private lands adjacent. My career took me a lot of different directions. . . . And then the job at Moose Creek [Ranger Station] opened up. I was offered the job and I jumped at it, and I guess the rest is history. I went out there in 1975, and we were well along in the process of doing fire management in a different way. The natural fire program. Rather than putting every fire out, we were allowing some of the fires to burn. 1979 was the first year that we fully implemented it to see whether it was going to work or not. We had seventy-five fires that year. It was a big fire year. We had the Independence Fire that burned about 18,000 acres. It was really the first test at Moose Creek of that program, and it worked. It was a good program, we saw the effects of natural fire where we didn't interfere with it.

LEE: Did it seem ironic to you that for fifty years the Forest Service had been fighting fires with all their might and implementing the 10:00 a.m. policy [a rule stating that Forest Service crews had to suppress fires by 10:00 a.m. the day after they were reported], and all of a sudden that was gone and now you were letting fires burn?

SEAMANS: Not really. Wilderness is a different breed of cat because in wilderness you don't have a lot of tools to manage vegetation, manage the forest. Outside of the wilderness we have a lot of tools . . . we can harvest timber, we can manage the forest. That [10:00 a.m.] policy of course derived out of the 1910 fires and the 1934 Pete King Fire. In wilderness, it wasn't a good policy. Outside of wilderness it had merits. So, I guess I don't feel badly about it, but I'm glad to see we are managing fire in wilderness because if it's to be affected by the forces of nature, one of the main forces of nature is fire. So yes, I think it was a good thing that we got the fire back into the wilderness ecosystem, and I think it was very appropriate that one of the first places we did it was the Selway-Bitterroot.

LEE: Now why do you think the Selway-Bitterroot was the place where natural fire management got its start?

SEAMANS: Well, it was one of the original wilderness areas in the 1964 Act and we were trying to be a leader in the country. In fact, I think we were recognized as a leader there. The rangers that were there before me and after me continued to try to make Moose Creek a leader in wilderness management in the country.

LEE: Can you think of some more areas in which Moose Creek was a leader besides fire?

SEAMANS: Well, in our dealings with the public. In the wilderness system, I think a lot of folks looked at the public as the biggest danger to the wilderness because we were trying to keep it as pristine as we could, and for all those natural forces to prevail. But we went into a program where we were trying to befriend the public. We weren't trying to bring use into the wilderness, but once people made a decision that they were going to visit the wilderness, we wanted them to have all the tools they needed to do it with as minimal an impact as possible. So, we had an education program going then, we had wilderness rangers that were in the field . . .

In the fall of the year, for example, we had a lot of hunters. Hunters were the primary users in the fall, backpackers in the summer and to some extent floaters on the river. . . . A maximum of opportunity and a minimum of regulation [for the users] is what we were after. Unconfined. And nothing is more confining than regulations. So, we tried to keep the regulations to a minimum. The use levels were low enough that people could travel at their own pace. They didn't have to stay at a particular camp ground. If they wanted to do the trip in three days they could. If they wanted to do it in ten days they could. 'Cause it was, again, unconfined.

With few regulations in such a large wilderness, people can get into trouble quickly, and in that case, human intervention is needed. Like all wilderness rangers at Moose Creek Ranger Station, Art experienced his share of search and rescue missions.

SEAMANS: We had an incident on the [Selway] river where our river ranger encountered a party on the river right up by, I guess he caught up with them not long after they left Paradise [the river put-in spot]. [The river ranger] was concerned about the quality of equipment they had for that river and the skill level they had. He went through Ham Rapid and showed them how to run Ham. They made it through Ham and they got down to Moose Creek and he said, "Well why don't you come down with me and we'll scout out Double Drop and Little Niagara and Ladle?"[2] And the guy that was in charge of the party, he said, "No, we don't have time for that we're gonna have to be down at Selway Falls by a certain time." So, they went without scouting it, and the river ranger he told me he was really concerned about this party. Well they got into, I think it was Little Niagara and they filled up with water and two people were thrown out of the boat, an oar was broken, and they drowned in Ladle. So, we had to recover those two bodies. The one victim was a teenager and the other was his grandfather. So, we had a couple there at Moose Creek, they'd lost their father and their son.

LEE: So how do you search for a body like that?

SEAMANS: Well, we used a helicopter in that case and were able to recover them the next day because they did have on life jackets. So, the life jackets helped us find the bodies.

In 1979, Art faced one of the most haunting moments of his career. On June 11, 1979, a Forest Service DC-3 airplane carrying twelve people into the wilderness for a summer orientation crashed into the Selway River near Moose Creek Ranger Station. Art was at the station preparing for the orientation when it happened.

SEAMANS: [Before the trip to Moose Creek Ranger Station] I'd made arrangements for an aircraft and I was thinking in terms of a Twin Otter, but there wasn't a Twin Otter available. They had a DC-3 that they said would do the same job cheaper so we brought the DC-3 up from southern Idaho. We had nineteen people on the manifest. I was going to fly in people first and then supplies. We had several trips planned for supplies, including horse feed and some lumber and other things that we needed to get back there. I went into Moose Creek to make sure everything was ready. We had what we called our guard school, which was really an orientation week where we covered tool use and fire control procedures and got people ready to do their jobs [for the summer]. It was the only time we had the whole crew together all year because after the guard school, they scattered around to the portals [small ranger stations at trailheads]. We had three trail crews that were backpack crews and then Emil and Penny Keck [two long-time backcountry workers at Moose Creek Ranger Station]. If there was any stock needed for heavy tools, they [Emil and Penny] did that work. So, we had our backpack trail crews in there and a couple lookouts. We had the Shissler and the Gardener lookouts.

Anyway, we'd have a get-together at the beginning of the season. We always had kind of a welcome party the first night, and had a barbeque and it was a nice thing to get us kicked off for the summer. And I had my wife and two girls were supposed to fly in, in that DC-3 that first trip. But we had a light plane coming in on the Friday before, and Joyce [my wife] decided she wanted to come in early so she could bake a cake and get some things ready for the party. So, she came in with the two girls so that's why they weren't on the flight. I had them on the manifest, but they weren't on the flight. . . . Ended up we had eight of our crew, plus two pilots that came in, and then two folks from the supervisor's office, the engineers.

'Course the airplane got a little over halfway and they had engine problems. They had to shut one engine down and when they did, they blew a piston that went down through the oil cooler and the prop seized, and when it did, it torqued the engine right out of the wing and that made the plane about as dirty as it can get, 'cause the wheels fell down and they broke a hydraulic line and their glide path was pretty steep. I'm sure the pilots were aiming for the one straight piece of the Selway River that they could've got the plane down on. The water was pretty high in the Selway, so there was quite a bit of water, but they were about thirty feet too short, too low, and they took the top out of a big Douglas fir and sheared off about twenty feet of the one wing. When they hit that tree I'm sure the plane went almost nose first into the river. It traveled very little distance after it hit the tree, and the pilot and copilot were crushed in the very front of the aircraft. But in the very tail end of the aircraft there were three people that survived the crash and they were able to get out of there. It was set up like smokejumpers [set up a plane]. They had the gear down one side, and people down the other side, so of course all that gear came loose, so everything in there was like a projectile. The three that survived were in the very tail of the aircraft and they ended up on a big boulder just off the south side of the river.

I don't know if you're interested in that story or not?

LEE: I'm on the edge of my chair, so anyway—

SEAMANS: Well, at any rate, I was getting ready for a class I was going to teach and one of my wilderness rangers came in and said, "Art, we got a situation." I went in and talked with the dispatcher out of Grangeville, and he said that the plane was overdue for a check in. They hadn't checked in as they were supposed to. So, we tried to contact them, had no luck and so I told him let's get an aircraft on the way in here and get a helicopter coming in from Fenn [Ranger Station, the closest station to Moose Creek] and he said, "They're already in the air." So, he was ahead of me. They found the ELT beeping and—

LEE: What's that?

SEAMANS: E-L-T. Emergency Location Transmitter. And the helicopter found the tail section of the aircraft. It had floated against the bank on the south side, and then he spotted the three survivors on this big boulder. Well, we also had an airplane with smokejumpers coming in from Missoula [Montana], and they jumped, one on one side and one on the other side of the river, and they had a crash kit with them with a pressure suit. It was a pretty complete kit, and the EMTs and the helicopter came up and picked me up and we went back down the river to the crash site.

I could see the three people on the rock, but we couldn't get to them with the helicopter because we didn't have enough clearance because of big trees. Big cedars. And the boulder was pretty close to the bank. So, we weren't able to get to them and that was very frustrating. We hovered out there on the river and one of the people was Andy Taylor. He was the engineering tech that I mentioned that I knew quite well, and Andy waved at me. One was a student aide and I believe he was sixteen, Bryant Stringham. The third was a young man that I had hired to work the portal at Selway Falls, and I'd talked with him on the phone, but I hadn't met him.

There were three [on the boulder] when I came up [river in the helicopter]. When I went down there were two. The smokejumpers were almost there. They were 30 feet above the boulder, and we had to pull out. We went down river and we were looking to see if we could find any other survivors, and we found some more wreckage down there. The wing section of the aircraft had lodged in Wolf Creek Rapid.

We went back to Moose Creek and as it turned out, well, I guess it was a few hours later, we were in the office there at Moose Creek when Bryant Stringham came walking straight into the station leading a horse, and of course we'd thought he was dead. We had a list of who'd survived, and we were desperately trying to get a big helicopter in from Spokane from the Air Force base up there that was capable of dropping a cable down to where they could lift those people out, and we were also trying to get a raft in there that would handle that water in the Selway, and our patrol boat. Our raft was enrouted to the launch site. And we got a BLM raft in from Cottonwood [Idaho] and we flew him in, but we lifted the survivors out with the helicopter. We had one survivor at that time [Bryant], because Andy Taylor had died.

LEE: The young man who waved at you? Did he die on the rock there?

SEAMANS: Yeah, about ten minutes after he'd waved to me. He had so much internal injury. Yeah, he bled to death. And they couldn't do a thing. They put him in a pressure suit but couldn't bring his blood pressure up at all, and he just passed away within minutes.

LEE: So, how do you get over a tragedy like that?

SEAMANS: You don't. You don't. And, then, the next few days, we were involved in dealing with parents and relatives and the investigation, trying to recover bodies and, I was involved in recovering three, four, five of the bodies.

LEE: And did you know them personally?

SEAMANS: Yeah, yeah, I did. Yeah, all but the Shissler Lookout, [who] was a young woman and I didn't know her other than having visited with her on the phone at some length. And her body was the first one we recovered. We were in two helicopters flying down the river. I was in the second one and Dale Swee [another wilderness worker] was a spotter in the first helicopter, and Dale called us on the radio and he said I've spotted one of the bodies and he said, "I think we can get it." The helicopter that he was in landed in a gravel bar in the middle of the river. It was covered with maybe a foot of water so they actually landed in about a foot of water. We came in behind them and landed and Dale had a hold, had a grip on the body, but he was in water up to his chest, and trying to keep a hold, and not lose the grip, so I went out and the two of us managed to get her back to the helicopter.

And there were the other bodies I was involved with recovering. We were hovering [in the helicopter] over an eddy where the current was coming down and around and the water was high, but it was clear. And I could see my saddle, and I had some lawn chairs that the jumpers had loaded on the plane because we had all of our stuff that we're going to fly in there for the summer, and so I could see that stuff floating around, but we couldn't see any bodies. As we were getting ready to leave—there were two of us spotting, and the pilot—and I said wait a minute, and I thought I'd seen something. We came back around and hovered there, and this body began to kind of... The sand was moving in the bottom of that eddy and the body just kind of emerged. Pretty soon, there were three that we could see there, and then the sand covered them again. We had divers, and we took the divers to the site and they put all the weights they could on one diver and sent him down. The water looked to be about ten feet deep, but it was actually about twenty feet deep, and the currents were so strong. The diver said he'd never seen anything like that. The sand was moving in such a way that he experienced the same thing we had. The sand, just... These things would rise out of the sand in the bottom of the river and then— *[Art starts to choke up]*

LEE: And then disappear again.

SEAMANS: And disappear. We were able to pull out three bodies there. And then the rest of the body recovery, I was not directly involved in. The divers... The pilot and copilot, we couldn't find. Their bodies weren't found until July. A fisherman found one of them, and some floaters found another. They were hidden under some overhanging rock.

Commentary

On April 13, 2013, I arrived in Lewiston, a town that sits at the confluence of the Snake and Clearwater rivers in northern Idaho. I parked on a residential street in front of a bungalow. Although I had conducted some fifty interviews over a three-year period for the Selway-Bitterroot Wilderness History Project, I never got over being nervous when I met someone for the first time. Art Seamans answered the door. A large man, he seemed imposing to me, which added to my unease. I knew that Art had spent most of his life in Idaho's wild lands, but his home seemed anything but wild. Photos of grandchildren hung on the walls. Well-tended plants sat on shelves. The modest furniture had a comfortable, lived-in feel. Joyce, his wife, made tea

Moose Creek Ranger Station Landing Field photographed from an airplane, May 1936. When Art Seamans came to work at Moose Creek in 1974, the closest dirt road was over fifty miles away. The only ways to access the station were on foot, by horse and mule, or backcountry aircraft. In 1964, with the passing of the Wilderness Act, the Selway-Bitterroot and Moose Creek Ranger District in particular became a model throughout the country and the world for a progressive wilderness management style that promoted the "free play of natural forces." *Photograph by George Case, courtesy of Shirley Wagoner*

and then retreated to a recliner in the living room while Art and I sat at the dining table. I began with standard questions about Art's background, and he answered in soft, low tones. I worried that the formality of the interview, such as the recording device and the prepared questions, would make the conversation stilted. But I was wrong. Art's stories unspooled into the long, ragged afternoon, and I relaxed into them. At the end, he took us both to an unexpected place. Art's interview was the last one I would conduct for the project, and, after I left, I thought about why oral history was the best method for documenting wilderness.

Early in the interview Art said his most challenging task as ranger in the Selway-Bitterroot was making sure "natural forces prevail." When he spoke of "natural forces" in this way, I did a double-take, for I knew these were the words of the author of the Wilderness Act, Howard Zahniser. While Zahniser was drafting the Act between 1956 and 1964, trying to get the wording just right, he wrote to a friend saying that wilderness should not be "subject to human controls and manipulations that hamper the *free play of natural forces*" (my emphasis).[3] The "free play of natural forces," he wrote, was best described by the single word "untrammeled." In fact, Zahniser's friends tried to convince him to use a different word, but he refused, and it was that word he eventually wrote in the Act's most oft-quoted paragraph: "A wilderness, in contrast with those areas where man and his own works dominate the landscape, is hereby recognized as an area where the earth and its community of life are untrammeled by man."[4] Untrammeled is an odd word, poetic in its richness, not a word used in policy documents. "Untrammeled" is best defined by its opposite. A "trammel" is an intricate, three layered net used for snaring birds and fish. More generally, "to trammel" refers to anything that restrains, controls, or confines. Untrammeled, then, suggests exactly the kind of free play Zahniser was after for a wide-reaching wilderness policy.

I had heard many of my narrators use "untrammeled," but none except Art referred to the free play of natural forces. In fact, he mentioned the word several times. I wondered if he was subconsciously drawing on Zahniser's letters, which few people knew well. I was never able to ask him, but after the interview ended, we talked off-record, where I found him conversant in politics, history, and ideas. He told me that before moving to Idaho as a young man, he had earned a graduate degree in forestry at Yale, the program set up by the first chief of the Forest Service, Gifford Pinchot. Although Art was humble and not given to self-promotion, he was aware that the Selway-Bitterroot had played a key role in the Wilderness Act and that he was part of that legacy.

Yet, that legacy began long before anyone conceived of a federal policy. The Selway-Bitterroot, which now comprises 1.3 million acres, drapes across the Idaho–Montana border in the northern Rockies. The land rises and falls in steep granite peaks and deep valleys cut by water and wind and covered in pine and fir trees. Humans have always interacted with the land, but in many ways the land has protected itself from human impact. The Nimiipuu people traveled the area for thousands of years, and it remains important to them as home to some of their ancestral and sacred sites, but by and large, the winters were too harsh for many permanent settlements. The land is shaped by a geologic formation called the Idaho Batholith, which is barren of minerals and produces soil as fine as sand or flour. These factors made the land unappealing to farmers, ranchers, and miners in the early twentieth century. As forester Franklin Reed wrote in 1925, the Selway-Bitterroot included "some of the roughest, rockiest country in America" with "no resources except scenery."[5]

The perception that the Selway-Bitterroot was not much more than pretty decoration kept it relatively free of roads and people until the 1930s, when wilderness advocacy came into its own. But at about that time, there was also a push to build logging roads. However, Forest Supervisor Guy Brandborg resisted that pressure. In fact, Brandborg openly criticized Forest Service practices that were in his view "forestry gone mad." He did so through passionate rhetoric, predicting ecological disaster as a result of the careless decisions he saw all around him: "They're wiping out animal habitats. They're scraping logging roads out of steep slopes where the gashes in the soil pour mud in the streams," he wrote.[6] Brandborg's vision was to keep the land as wild as possible, a vision that caught the attention of the now-famous conservationist Bob Marshall. Much like Art Seamans, Marshall had moved as a young man from the east coast to Idaho while earning his graduate degree in forestry at a prestigious school—the Johns Hopkins University.

And just as Art would, Marshall fell in love with the American West, and the Selway-Bitterroot in particular. Wonderfully articulate, Marshall wrote about the wilderness in dozens of documents, including a 1934 letter to the Interior Department. Having just finished a long hiking trip through the region, he was full of mountain scents and river sounds and hell-bent on setting aside Idaho's land, which he did in a style that combined hard logic, poetic observation, and humor:

> This is an era of planning. Cities are planned, factories are planned, the production of wheat, beans, and, I presume, artichokes is planned, currency and wages are planned, and the construction of highways is not only

> planned, but also platted glowingly on gasoline company road maps. I have in my hand one of these latter for the state of Idaho, and the depressing, snake-like meanderings in red and blue across areas I not so long ago tramped while sixty miles from the closest possibility of mechanized transportation, reminds me that the one thing conceivable which doesn't seem to be planned, is the preservation of natural conditions.[7]

Planning for the preservation of the natural world became Marshall's life work. In 1937, after another long hike in the Selway-Bitterroot, he visited Guy Brandborg to talk about a wilderness system that could span the entire country. Brandborg's son Stewart was only twelve at the time, but he remembered vividly the visit because of Marshall's badly sunburned face. Young Stewart could not have foreseen that Marshall would die shortly after that visit, and that he, Stewart, would grow up to become a conservationist in his own right. Not only that, but in 1956 Stewart accepted a job in the Wilderness Society, the very organization Marshall had co-founded, and in this position Stewart helped Howard Zahniser write and rewrite the Wilderness Act sixty-six times and shepherd it through Congress. Thus, the Selway-Bitterroot, a favorite of Bob Marshall and home to Guy and Stewart Brandborg, played a key role in developing the concept of wilderness in the United States, a definition subsequently adopted in other countries. Not surprisingly, the land was among the first regions to be protected in 1964.

Art Seamans began working in the region in 1961, and by 1974 he was a Selway-Bitterroot Wilderness ranger heading up Moose Creek Ranger Station, one of the most remote stations in the country. Here, he was responsible for implementing the tenets of the Wilderness Act. Thanks to Art and others like him, the Selway-Bitterroot remains to this day one of the most untrammeled places in the lower forty-eight states, a sentiment I heard many of my narrators express. Rob Mason, who served as director of the Selway-Bitterroot Foundation when I interviewed him in 2011, said: "The thing that makes the Selway [Bitterroot Wilderness] special is it is super wild. I mean, it is the epitome of wilderness . . . because it's so big, and there are so few intrusions from the outside . . . and also the populations around it are low, and so you don't get that many people there, and it's so rugged. I mean it's unquestionably the most rugged place I've ever been."[8]

Oral history, I came to realize during discussions with people like Rob and Art, is one of the few methods that can capture the spirit *and* history of a wild place. Although oral history is a fact of human culture, it came into its own the early twentieth century and caught fire in the 1960s, following a similar timeline to the Wilderness Act.[9] It is perhaps no coincidence that just as wilderness honors untrammeled land, oral history method privileges untrammeled narrative. At its inception, for example, oral history method drew on anthropology, but departed from some of its more rigid techniques. Instead of observing people in other cultures with supposed objectivity, oral historians listened to their subjects recall the past from memory, and they often became participants in the meaning-making process. This methodology was not only a sharp turn from anthropology, it also diverged from that of traditional historians who sought precision in chronology,[10] whereas oral historians by necessity sought freedom from linear time.[11] Moreover, traditional historians valued fixed forms—documents, court papers, log books, and the like—over the fluidity of oral narrative.

That fluidity is oral history's strength, especially when documenting wilderness, since oral history lives in the space between human agency and natural forces. Take sound, for example, the primary sense evoked in oral historiography.[12] Spoken phrases, sentences, and paragraphs are made of the air that moves between people, and these cannot be captured except on recording devices.[13] Sound, according to ethnobiologist David Abram, is nothing more than "shaped breath" and just as ethereal.[14] This air or breath is one of nature's wild forces. Ecopoet Gary Snyder defines "wildness" as any process not directed by human agency, such as "the involuntary quick turn of the head at a shout, the vertigo at looking off a precipice, the heart-in-the-throat in a moment of danger, the catch of the breath, the quiet moments relaxing, staring, reflecting."[15] In this definition, our memories, imaginations, and the novel sentences we speak aloud extemporaneously are wild, too.[16]

Furthermore, sound is often inseparable from place, a crucial factor when one is chronicling wilderness, since oral history allows narrators to give voice to natural forces.[17] I discovered that for narrators who had spent years in the wilderness, sound extended beyond human speech to the land itself. They had carried land sounds out of the wild and into the domesticated interview sessions. They sang to me in the voice of a Swainson's thrush, mimicked the gurgle or roar of a creek, imitated a bear's growl, or simulated a mosquito whine. One narrator, the former Selway-Bitterroot ranger and photographer Dick Walker, said, "I have photos where the ptarmigan are bursting out of the snow. And then they go under the snow and they're warm, see. So, you're skiing along, and all of a sudden, *poooof!*"[18]

Perhaps without realizing it, my narrators were part of a legacy linking sound and wilderness. The sounds of the land had been vital to the traditional oral cultures in the Selway-Bitterroot for thousands of years. Elsewhere, I have written about how the Nimiipuu heard the land and its animals speak.[19] Coyotes, fish, mountains, wind, and rivers had voices. During my research, I came across a 1970 recording of the Nimiipuu elder Elizabeth Wilson, an important Nimiipuu historian. One of the most astounding passages on that recording was a conversation between Elizabeth and her son Angus in which the two talked about the wind blowing through a dead snag on the river. As they discussed the sound, they made a lonely, moaning noise. "You only hear it in the mountains where fire has swept through," Angus said. "And even one snag will make noise. But where there's a bunch of them, you just hear—*woooaaaanh-woooaaaanh*—it sounds like a band of sheep." For indigenous people and wilderness workers alike, sounds and spoken words are the free play of Earth's natural forces, as demonstrated by the Wilsons' moaning and Dick Walker's *poooooof!*

Sound may be one of the most pervasive natural forces, but fire is one of the most powerful. As Art put it, if wilderness is "to be affected by the forces of nature, one of the main forces of nature is fire." Prior to 1964, lightning fires were suppressed because the public believed fire was destructive. But with the passing of the Wilderness Act, the Forest Service began to see fire as regenerative. In 1972, just before Art became a ranger, the chief of the Forest Service approved an experimental plan to let a wildfire burn without human intervention. He chose the fire on White Cap Creek in the Selway-Bitterroot Wilderness to set the experiment because it was so remote.[20] As fire ecologists predicted, the White Cap Creek Fire burned out naturally, scorching some areas and jumping over others completely. It stopped at natural boundaries such as a body of water or a fire-scarred patch grown back in young plants. Fire scientist Robert Mutch helped supervise the White Cap experiment and later implemented the "Let Burn" policy throughout the country and the world. During my 2011 interview with Mutch, he described burned land as a mosaic: "Some places burned the crowns of the trees but it didn't necessarily kill all the crown, and there were places where it just burned as an understory fire and you couldn't even see signs of it because just the surface fuels, the pine needles, branches on the ground burned, and there were places that were skipped entirely in that fire. So, it leaves a very interesting landscape afterward."[21] As he spoke and showed me photographs of various burns, I understood how unregulated fire left text on the landscape.

"Free texts" is what oral history pioneer Jan Vansina called oral testimonies (as opposed to memorized oral forms such as prayers, ritual messages, poems, and epics).[22] Although he did not have fire or wilderness legislation in mind, the freedom of oral story mirrors that of fire. Like fire, oral testimonies, when not suppressed, move through memory in a nonlinear, mosaic, or "hopscotch" pattern, as oral historian Barbara Allen terms it.[23] Oral story crowns at certain points, contradicts and burns back at others, and sometimes jumps over patches of memory altogether, not touching them at all. Indeed, if fire is the most salient example of the free play of natural forces in wilderness, memory is the most important kind of free play in oral history methodology.

The biochemistry of memory is still poorly understood. Recent scientific research even suggests that after memories are constituted, they are destabilized and reconsolidated again with each recall of an event.[24] Some oral historians are trying to unravel the subtle relationship between memory and recall in interviews. Kenneth Kirby, for example, summons early twentieth-century phenomenologist Edmund Husserl, who saw memory as a complex ground in continuous flux.[25] For Kirby, this flux reflects the way we process conscious life and store it later, as well as the fluidity oral historians encounter when they ask narrators to dig deep into their memories. Such ebb and flow can be frustrating when narrators do not recall events or details of their experiences, but it can also be the most exciting and surprising part of an interview when they do, a wilderness unto itself.

The play between memory and speech is just as fluid. Long before oral history was a subject of study, William James explained the mysterious process by which thought becomes speech. In "Stream of Thought," a chapter in his *Principles of Psychology* (1890), he marvels at how in speech, human beings start a sentence without knowing exactly where it will lead. We rush "headlong" into a sentence, gather intention and energy, and bring it to a conclusion before we know what we have said. If we happen to separate our thoughts from the sentence we are speaking, it ceases to be itself. It is like stopping a river to try to catch its movement.[26] Oral history methodology must be open to memory's free play not least because on some occasions, narrators recall what they have never before put into words.

Memory is an especially important tool in documenting designated wilderness because such spaces live partly in human memory. Connie Saylor worked as a wilderness ranger in the Selway-Bitterroot for twenty years. After she retired, she yearned for the land, which cannot be privately owned (with the exception of a few inholdings that were grandfathered in with the Wilderness Act). In fact, people can visit wildernesses, but they

A helicopter pulls debris from the June 11, 1979 DC-3 aircraft crash in the Selway-Bitterroot Wilderness near Moose Creek Ranger Station. Art Seamans was at the station when he heard the news of the crash. He spent days hovering above the river during the recue, body recovery, and clean up. *Photograph by Art Seamans, courtesy of Cindy Schacher*

can only stay there for three weeks at a time. When I interviewed Connie, she told me the only way she could possess the land was by "owning it in my heart," by which she meant the intensely vivid memories she carried with her after she left.[27] I was in fact startled by how my narrators' memories assumed different forms. One woman, the textile artist Sarah Swett, construed the wilderness literally as an image. As a young woman of twenty, she had worked in the Selway-Bitterroot in the 1970s, and later, in her fifties, she began weaving what she remembered into large tapestries. Ed Bloedel was on the regional wilderness staff on the Selway-Bitterroot in the 1960s. When I met him, he recalled the land through song, playing homespun compositions on his guitar.

For Art Seamans, the wild Selway River was the site of memory, especially the traumatic DC-3 crash on June 11, 1979, which he spoke of toward the end of the interview. Compelled by his story, I wondered if oral history method had played a part in Art's recollection of the event. A few years after the interview, I compared my 2013 interview transcript with an article about the crash in the *Lewiston Tribune* of June 14, 2009 in which Art was interviewed.[28] I found that indeed, Art repeated several narrative elements in the article and the oral history. He told the *Tribune* reporter: "I was preparing some lesson plans I was going to be teaching." Similarly, he told me, "I was getting ready for a class I was going to teach." The echo of this particular narrative element made sense. Art choreographed the moment when he first heard something was wrong with the plane, a moment of recognition, a technique any good storyteller would use. In fact, Art was a particularly skillful storyteller, using restraint and understatement to heighten narrative effect. But his real power came from a different place, one of honesty and vulnerability. In both the newspaper and the oral history interview, he also narrated the moment when he realized his wife and daughter were supposed to be on the fated DC-3 flight but had actually flown into the wilderness the previous day. "I had them on the manifest, but they weren't on the flight," he told me, obviously still haunted by the tragedy and grief, and perhaps even shadowed by guilt that his family had survived while others had not.

Not surprisingly, though, the oral history session, with is emphasis on memory's free play, went beyond these narrative points to something deeper, something Art had not talked about before: the mystery of recovering the bodies. His daughter confided in me later, "Dad didn't talk about the crash or its aftermath very much. That day weighed heavy on him for the rest of his life. He once told mom and I that there wasn't a day that went by that he didn't think about it and those that lost their lives."[29] In other words, the oral history process—the tape recorder, meeting in his home over tea, myself as the interested listener "on the edge of my seat," even the formality of signing the IRB forms—rather than stilting the interview or putting a barrier between us actually brought us closer. Art's willingness to delve into an emotionally haunting story is similar to what ethnographer Ivo Strecker found when he recorded the Hamar people in Ethiopia. Instead of clamming up or becoming timid, people would open up, offer more of themselves, and become more generous and expressive.[30]

At the end of Art's interview, his narrative took an even more radical turn, one that caught me by surprise and that I believe the oral history process facilitated. In his discussion of the body recovery, he spoke not just of the river as a site of memory, but as if it were memory itself. "We came back around and hovered there, and this body began to kind of . . . The sand was moving in the bottom of that eddy and the body just kind of emerged," he said. Art choked up as he told about how, in twenty feet of water under strong currents, the sand moved at the bottom of the eddy and released the bodies so they emerged briefly before sinking back down. Like the psyche itself, the river disclosed its secrets according to its own natural laws: currents, the amount of water rushing through rocky channels, the play of sand along the bottom. "These things would rise out of the sand in the bottom of the river and then . . . disappear" with a fluidity resembling the mysteries of the human mind in reflection.

When I listened to Art's recording a few years later, I was especially struck by his use of the word "hover." "We hovered out there on the river," he said as he talked about being in the helicopter in the rugged terrain. Without diminishing how difficult it was for him to execute the body recovery of people he knew well, I thought about his repeated use of the term "hover." There's some evidence that scientists, musicians, and artists—from the biochemist Sir Hans Krebs to the poet Samuel Taylor Coleridge to Mozart—have used that term to describe what happens in mind and memory in the creative process.[31] These people "hover" over a subject, not in a systematic way but through free play. They keep a topic in mind while doing other activities like cooking a meal, walking, or dreaming, until the answer reveals itself. Whether in a helicopter over a wild river or in the far reaches of the human mind, hovering discloses what is submerged.

I am grateful that the free play of oral history method allowed Art access to his own deep memory about the DC-3 crash. His daughter said that the memory "weighed heavy on him." Since my interview, the story has weighed heavily on me, too, for it shows us what we can hear if we listen closely, and yet how much remains buried. "The pilot and copilot," Art said, "their bodies weren't found until July. A fisherman found one of them, and some floaters found another. They were hidden under some overhanging rock." Art taught me through his oral history that wilderness reveals and confides its secrets in its own way, and in its own time.

Notes

1. Art Seamans passed away on June 12, 2014.
2. According to the organization American Whitewater, "the Selway River is a nationally designated Wild and Scenic River flowing through the Selway-Bitterroot Wilderness Area. As a major tributary of the Clearwater, it was one of the original eight Wild and Scenic Rivers designated in 1968. The Selway is one of North America's premiere wilderness whitewater trips where boaters typically spend three to five days on this forty-seven mile run." Moose Creek joins the Selway about half way down the river, significantly increasing the Selway River's volume. This inflow is known in the white-water community as "Moose Juice." From Moose Creek Ranger Station, there are four miles of class four and five rapids including Double Drop, Ladle, and Little Niagara. https://www.americanwhitewater.org/content/River/detail/id/623
3. Doug Scott, *The Enduring Wilderness: Protecting Our Natural Heritage through the Wilderness Act* (Golden, CO: Fulcrum, 2004), 2. For a more recent discussion about the word "untrammeled," see John Copeland Nagel, *Law's Environment: How the Law Shapes the Places We Live* (New Haven, CT: Yale University Press, 2010).
4. The Wilderness Act of 1964. Public Law 88-577 (16 U.S.C. 1131-1136), 88th Congress, Second Session, September 3, 1964. Wilderness.net. https://www.wilderness.net/nwps/legisact
5. Franklin Reed wrote in response to conservationist Aldo Leopold's 1925 discussions of wilderness. AnneMarie Moore and Dennis Baird, eds., *Wild Places Preserved: The Story of Bob Marshall in Idaho* (Moscow: University of Idaho Library, 2009), 43.
6. Neal R. Peirce, *The Mountain States of America: People, Politics, and Power in the Eight Rocky Mountain States* (New York: Norton, 1972), 106–7. David Gerard, "The Origins of the Federal Wilderness System," in *Political Environmentalism*, ed. T. L. Anderson (Stanford, CA: Hoover Institution Press, 2000), 214.
7. Moore and Baird, *Wild Places Preserved*, 83.
8. Rob Mason, interviewed by Debbie Lee, Moscow, Idaho, March 5, 2011, Selway-Bitterroot History Project, Wilderness Archives, University of Idaho, Moscow.
9. Although oral history method and designated wilderness share timelines and privilege the unplanned to the planned, free play to strict rules, there are relatively few oral history projects devoted to wilderness issues. Some that do are the Sierra Club Oral History Series in the Bancroft Library at UC Berkeley featuring conservationists like Ansel Adams and Wallace Stegner. The archive covers preservation and sustainability topics more broadly and often addresses wilderness management. Likewise,

"I Remember Yosemite," a multi-year oral history project that records the stories of people who have helped shaped that iconic area, covers wilderness sections in the Yosemite National Park. And there is also Project Jukebox, developed by the University of Alaska Fairbanks Oral History Program. While these projects often touch on wild lands, none are devoted to exclusively to one wilderness area. In fact, the Selway-Bitterroot Wilderness History Project is the only one to account for the human history of a single wilderness. The goal of the project has been to document the voices and ideas that shaped such a large and pristine landscape and by doing so become a model for other scholars.

10. Gwyn Prins, "Oral History," in *New Perspectives on Historical Writing*, ed. Peter Burke, 2nd ed., (University Park: Penn State University Press, 2001), 125.
11. See Jan Vansina, *Oral Tradition: A Study in Historical Methodology*, trans. H. M. Wright (New Brunswick, NJ: Transaction Publishers, 1961, 2009).
12. Paula Hamilton, "The Proust Effect: Oral History and the Senses," in *The Oxford Handbook of Oral History*, ed. Donald A. Ritchie (New York: Oxford University Press, 2011), 224. "Since oral histories depend for their existence on the technology of sound, it is little wonder that hearing the past features prominently in interviewing. Some oral historians feel that sound should be taken far more seriously as the central medium of communicating memory" (224). Also see Steph Ceraso, "(Re)Educating the Senses: Multimodal Listening, Bodily Learning, and the Composition of Sonic Experiences," *College English* 77, no. 2 (November 2014): 102–23.
13. David Abram, *The Spell of the Sensuous: Perception and Language in a More-Than-Human World* (New York: Knopf Doubleday, 2012), 224–5.
14. Ibid, 180.
15. Gary Snyder, *The Practice of the Wild* (Berkeley, CA: Counterpoint, 2003), 16.
16. William Cronon, picking up Snyder's idea, argues "wildness (as opposed to wilderness) can be found anywhere." See Cronon, "The Trouble with Wilderness; or, Getting Back to the Wrong Nature," in *Uncommon Ground*, ed. William Cronon (New York: Norton, 1996), 89.
17. Gavin J. Andrews et al., "'Their finest hour': older people, oral histories, and the historical geography of social life," *Social & Cultural Geography* 7, no. 2 (2006): 7.
18. Author's field notes, April 25, 2009, Peck, Idaho.
19. Debbie Lee, "Listening to the Land: The Selway-Bitterroot Wilderness as Oral History," *The Oral History Review* 37, no. 2 (Summer/Fall 2010), 235–48.
20. "The Selway-Bitterroot Wilderness Fire Program," Cohesive Wildland Fire Management Strategy National Goals; Collective Solution: Response to Wildfire Adapted Communities Resilient Landscapes Supported by Science. Forests and Rangelands Website. https://www.forestsandrangelands.gov/documents/strategy/rsc/west/stories/WRSC-SBW-Success.pdf
21. Robert Mutch, interviewed by Debbie Lee, Missoula, Montana, September 13, 2011, Selway- Bitterroot History Project, Wilderness Archives, University of Idaho, Moscow.

22. "New Perspectives on Historical Writing," introduction to Jan Vansina, *Oral Tradition: A Study in Historical Methodology*, trans. H. M. Wright (New Brunswick, NJ: Transaction Publishers, 1961, 2009), 50.
23. Barbara Allen, "Re-creating the Past: The Narrator's Perspective in Oral History," *The Oral History Review* 12 (1984): 3.
24. Ana Paula Crestani et al. "Memory reconsolidation may be disrupted by a distractor stimulus presented during reactivation," *Scientific Reports* 5, article number 13633 (2015), http://doi.org/ 10.1038/srep13633.
25. Kenneth Kirby, "Phenomenology and the Problems of Oral History," *The Oral History Review* 35, no. 1 (Winter-Spring, 2008): 30.
26. William James, *Principles of Psychology in Two Volumes*, Volume 1 (New York: Henry Holt & Company, 1890), 224. James uses the example of "seizing a spinning top to catch its motion."
27. Connie Saylor, interviewed by Debbie Lee, Nez Perce, Idaho, May 6, 2011, Selway-Bitterroot History Project, Wilderness Archives, University of Idaho, Moscow.
28. Kathy Hedberg, "Forest tragedies haunt survivors: Moose Creek Ranger District was the site of two plane crashes separated by 20 years that cost many lives," *Lewiston Tribune*, June 14, 2009.
29. Email exchange with Cindy Schacher, Central Zone Archeologist, Lochsa, Powell and Moose Creek Ranger Districts, March 20, 2017.
30. Ivo Strecker, *Ethnographic Chiasmus: Essays on Culture, Conflict, and Rhetoric* (East Lansing: Michigan State University Press, 2011), 2–3.
31. Russell Meares, *The Metaphor of Play: Origin and Breakdown of Personal Being*, 3rd ed. (New York: Routledge, 2005), 54–5.

11

Culture Keepers

Voices of Renewal in the Eurasian Taiga

Tero Mustonen

Oral histories by Tero Mustonen with Evenk reindeer herder Vladimir Kolesov, Sakha-Yakutia, Siberia, Russia (2005–2010) and Finnish-Karelian fisherman-farmer Kauko Heiskanen and Elvi Heiskanen, North Karelia, Finland (2013) as part of the Snowchange Oral History Archive

The boreal taiga[1] *is a home to several rooted societies, which have preserved histories, endemic cosmologies, and a close proximity to natural ecosystem up until recent times. These oral histories arise from the efforts of the Snowchange Cooperative, based in Finland. Snowchange is a nongovernmental organization and community-led network across the Eurasian North whose goal is to preserve and renew taiga- and tundra-based traditional livelihoods, cultures, oral histories, and ways of life. The Snowchange Oral History Archive*[2] *preserves much of the work of the initiative, a repository of traditional knowledge materials from the member and partner communities.*

Two communities were involved in the co-production of the oral histories presented here. First, somewhat inspired by the work of the Snowchange Cooperative and the long-running Igloolik Inuit oral history work from Canada, the Institute of the Indigenous Peoples located in the regional capital of Yakutsk, Sakha-Yakutia, chose Yiengra as a site of oral history work to document changes of the environment and climate in 2004. Snowchange was asked to carry out this work in cooperation with the Evenk.[3]

Tero Mustonen, *Culture Keepers* In: *Oral History and the Environment*. Edited by Stephen M. Sloan and Mark Cave, Oxford University Press. © Oxford University Press 2022. DOI: 10.1093/oso/9780190684969.003.0012

Interviews with Vladimir Kolesov, Sakha-Yakuita, Siberia, Russia, 2005 to 2010

Vladimir Kolesov was a reindeer herder and hunter from the nomadic community of Gonam. I worked with him from 2005 to 2010 in the taiga camps of the 4th Brigade of the Reindeer-herding kolhoz, *or state farm, in Yiengra, the Republic of Sakha-Yakutia. We worked together right up until the time of Kolesov's untimely death, recording oral histories, visiting the nomadic camps, and discussing the role, scope, and meaning of the oral histories and cosmology that Vladimir still possessed. He was part of Clan "Buta" and belonged to the "Gonam" tribal indigenous community, or an* obschina, *which came into existence at the end of the Soviet era, post-1991.*[4] *The* obschiny *are concerned about their rights, which have been guaranteed by Russian federal indigenous legislation, laws that safeguard the indigenous culture, language, and land rights in Russia.*[5]

Having been immersed in the local reindeer herding and hunting culture from birth, Kolesov was one of the youngest speakers of the very nuanced and articulated Evenk language. He lived most of his life in the taiga forest. In April

Evenki reindeer herder and hunter Vladimir Kolesov surveys the mountainous landscape of the Yiengra river basin in the spring of 2010. Kolesov mastered the taiga skills of his people, as well as the cosmology and knowledge of a good life with the forest. *Photograph by author*

2005, as we were sitting around the fire in the tent, Kolesov slowly reflected on the culture of the Evenk customary law.

KOLESOV: We say that there is Mother Nature.
[There are strict rules] for behaving in the taiga.
If you cross a big river you need to tie a piece of cloth as an offering to a tree on the shore.
This applies to passing mountains too.
You should not throw firewood around—it should be in proper stacks.
You should not cut more than you will use.
You only should fish what you need—no more.
If you have no need—do not use or take.[6]

Vladimir's village settlement, Yiengra, is located in the southern part of the republic and it is home to 800 to 900 Evenk people. From 2006 to 2017, several industrial development projects connected to extensions of coal and gold mines, the establishment of several hydroelectric stations, and railroad connections and the East Siberia Pacific Ocean Pipeline greatly modified the lands and lives of the people in the region. The first railroad, the Baikal-Amur tracks (or the BAM track), had already been constructed across the taiga in the 1970s. Vladimir related that when the tracks were constructed, several young reindeer herders committed suicide. They could not come to terms with the imposed dramatic changes in their lands and lives. They could not find any use for their traditional skills and values, as the whole universe around them had been turned upside down. Compounding these changes brought to life in Yiengra is the rapidly proceeding climate change. One topic that the village and Evenk leadership, such as the esteemed philologist and folklorist Galina Varlamova, asked us to inquire about was the impact of the rapidly changing weather on life in the nomadic camps.

When asked how he sees the future, Kolesov, after much reflection, stated:

KOLESOV: Usually it was common to have minus 50 C [Celsius] in the wintertime in the past.
Sometimes it went down to minus 60 C when I was at school.
But now we witness climate warming.
It is rare nowadays to have even minus 45 C temperatures.

Last winter [2004–2005] was very warm, with only minus 30, minus 35 C at the coldest time.

I do not know why it is so.

Perhaps it results from the mining activities, industrialization of our region.

Warming is real, it is happening.

Weather has changed so much.

And the old markers to predict weather do not hold true anymore.

He continued to narrate reflections of the future.[7]

KOLESOV: Language will disappear.

I have great fears.

We are explaining all that we can to our children, but I am still afraid.

I am afraid even to speak about the things, which are coming.

We will try to preserve all that we can.

We will do all what we can to achieve that.

But several Nations have been wiped out from the face of the Earth.

We Evenk are a very small people.

I am afraid shamanism will die too in the similar fashion.

You need to have proper [taiga] teachings to be able to become a shaman.

You cannot just decide to become one . . .

There used to be epic singers who mastered the [mythical songs]. They are gone now.

Like the brother of my grandmother, Nikolai Germoginovitch Trafimov.

Perhaps after a while there will be no one left to discuss things in the same way as I can now.

If we are only drinking we are not following [customary rules].

Perhaps we can still fight.

Perhaps we can still prolong our existence.

If we would be further away from the road [and railway], language and culture would be preserved better.

. . . We are still nomadic; we are still on the taiga.

We still go on.[8]

I met Kolesov for the last time in the taiga in March 2010. On the last meeting, he wished me well. In early July 2010, I heard that he had taken his own life.

Interview with Elvi and Kauka Heiskanen, North Karelia, Finland, 2013

Fisherman-farmer Kauko Heiskanen from Eno, North Karelia, Finland has been working in the nature-based economies, such as fishing and small-scale farming of endemic species, since the 1940s. Through his life, by experimenting and revitalizing approaches building on traditional Finnish-Karelian practices, Heiskanen has emerged as a defender of the last remnants of the Eurasian taiga boreal old-growth forests of North Karelia and a strong advocate of intact ecosystems and place-based economies. Kauko and his wife, Elvi, are professional strawberry farmers living in the village of Kuusoja, in Joensuu, in North Karelia. In their lifetime, they have practiced a range of subsistence and traditional economies, including professional freshwater fisheries, the collection of mushrooms, and other seasonal trades. At the time of the interview in 2013, they cultivated local species, such as swidden turnip[9] *and potato onion,*[10] *traditional staples of the region since the advent of the endemic slash-and-burn agriculture that came with early farming in the communities. Their son Jari continues the farming tradition.*

MUSTONEN: How did you develop an interest in these native species of turnip and potato onions and ways of preserving these stocks?

K. HEISKANEN: We have been cultivating the potato onion for over forty years. People have really liked those onions we sold to customers, but now for some years we have not cultivated them for the commercial retail purposes. . . . It came from the homeyard of Elvi. . . . Then this turnip, I got the first seeds from an old grandfather there in the Kuusjärvi village during a local cultural festival. I think he was eighty years old. He had this *naurishauta*, a traditional oven on the ground for cooking the turnips he had there. He was very happy that we would be continuing preserving these stocks. I think we started in 1976 or so. I think it is very important to do this because old people found this to be a good stock and for hundreds of years they have been cultivated here, even before the arrival of potato.[11]

The oral history work commenced just as Areva, a French mining company announced plans to open a uranium mine near the home of the Heiskanen family. Fishing is a central traditional artisanal and subsistence practice in the region, and it was natural to document some of the elements of North Karelian fisheries with the Heiskanens as a part of the oral history work.

Kauko Heiskanen points with a stick to a spruce tree that he has partly clipped using his traditional knowledge. The reason for this procedure is to speed up the transformation of an economic forest lot back into a forest reflecting the boreal old-growth characteristics. The forest is emerging to be a good habitat, for example, for the Eurasian three-toed woodpecker, an indicator species for a diverse forest habitat. These acts of rewilding are essential to bring back the diversity of forest ecosystems in Finland. *Photograph by author*

MUSTONEN: You have been involved with Elvi also both on subsistence and commercial fisheries. Could you speak about that a little bit?

E. HEISKANEN: We were fishing for vendace on Lake Pielinen at spawning time. We also harvested fish using winter nets under the ice during the winter. That was so cold activity!

K. HEISKANEN: For vendace we usually fished for a month [at the time of the spawning] for vendace. Every morning we set out, no matter how strong the winds would be or how small the boat.

E. HEISKANEN: And what storms we experienced!!!

K. HEISKANEN: We had approximately fifteen to twenty nets out, they were five meters tall. We camped at nighttime in the islands during the fishery.

E. HEISKANEN: Yes, we overnighted at the islands. We were young so we were strong! We made fish soup and fried the fish and all of that. We did everything there.

K. HEISKANEN: I did all the nets and the fish traps myself. I had a friend, Mönkkönen, who made the nets himself, so I learned from him. Sometimes we also fished for common bream with fish traps during the spawning times. We really got good catches through that! But you need to pay close attention to the timing of the common bream spawning in the spring. Occasionally you could catch over one hundred kilograms at once in one trap, and then the trap was really almost tearing apart with the fish. You really need to know the places where they spawn. But now that the water levels are fluctuating [because of the hydroelectric stations] they do not always appear on the same spots for spawning.

MUSTONEN: What kind of thoughts did you get when the hydroelectric stations were constructed on the River Pielisjoki?

K. HEISKANEN: It really triggered bad feelings. Nobody asked the opinion of the local people. I remember when the Saapaskoski rapids used to be free. I went there as a small boy to fish with a rod. A land-locked salmon jumped all of a sudden. They also had these fish traps made from twigs for the salmon. Really well made. The salmon jumped into the trap. The traps were open and they had been positioned into the rapids so that salmon would jump in them. Traps were made from spruce or juniper. Different farms had their own harvesting places on the rapids. They could not overfish the salmon using those methods. These hydropower stations have cause irreversible damages for sure.

Although most of the oral history work focused on the various environmental drivers of change in the region, slowly we started to reach a deeper theme—how did Kauko become aware of an older relationship with the forest?

MUSTONEN: When and how did you become aware of "another relationship with the forest," one not connected with the present-day industrial logging?

K. HEISKANEN: Let us say I was doing forestry for about fifteen years, clearcutting old-growth forests. Then it occurred to me that this is not the right thing to do. I stopped doing that. For example, when I think of the Siberian flying squirrel[12] and the nests it has on these trees. It was almost like a pet. And all of their nesting trees were destroyed. It must be a problem from nature's viewpoint that all these forests have been lost. Some of the more sensitive species I am sure have become threatened, endangered. . . . In our forests, in the old growth there are these ant hills, under the spruce trees. They are very high. And so well-built! I think they construct their hills based on the weather conditions. If it is a rainy summer they make the top of the hill sharper, more pointier. But then close to a clear-cut they make it less steep, otherwise the hill might warm or dry too much. Most of the ants themselves will be underground. If these areas are clear-cut, the ants definitely move, they cannot stand forestry activities.

MUSTONEN: What is the root cause in your opinion, why are people still clear-cutting and destroying the forests? Is it money or what?

K. HEISKANEN: I think it is money overall. You need to acquire so much money at once. And make the forests productive, profitable.

MUSTONEN: What does nature mean to you?

K. HEISKANEN: It is everything in my life. I am even restoring some of the forests back towards old-growth, rewilding them. They will age faster. I can see some of the results already. For example the Eurasian three-toed woodpecker[13] has made its home in my forest. Last summer it already nested there.

Their oral histories were recorded with the Heiskanens throughout 2013 as a part of the "Cultural Heritage of the Vaara Karelia"[14] project to document and highlight the little-known traditions and oral histories of the region. Although influenced by elements from both Savo-Karelian and Karelian proper throughout the centuries, these ways of life and stories retained their own specific characteristics. In 2017, I contacted Kauko to discuss the latest idea of translating and presenting excerpts of these oral histories to international audiences. Kauko was very approving and enthusiastic about these ideas.

Commentary

Nature in North Karelia consists primarily of southern boreal taiga ecosystems with large lakes created by the last ice age. It is a part of the

Eurasian taiga zone. Eastern parts of the province have a more continental climate compared to western Finland, where the Baltic Sea provides a milder climate. The ecosystems, however, have witnessed massive human-induced damage since the 1960s.

The most significant threats to the biodiversity in North Karelia include the fragmentation of continuous ecosystems due to industrial forestry practices, which also contribute to the loss of marshland habitats, impacting on freshwater ecosystems. Up until the late 1800s, and even up to 1939 (at the onset of the Second World War in the region), the ecological systems in North Karelia remained in relatively pristine condition. The livelihoods of local people were mostly centered on small-scale slash-and-burn agriculture, hunting, and fishing.

After the Second World War, Finland had to pay war damages to the Soviet Union, which meant a rapid industrial revolution and intense use of natural resources also in North Karelia. Most old-growth forests were clear-cut. Marshlands were transformed into peat production. Waterways such as River Pielinen, home of the freshwater salmon (*salmo salar*), a relic of the Ice Age, were harnessed for electricity production. Farming shifted to chemical-driven mass production.

After World War II, the search for better opportunities in Finnish towns or emigrations to Sweden emptied many communities in the countryside. The pulp and paper industries polluted the waters. From 2006 to 2013, uranium and other large-scale natural resource extractions were being planned in the region, offering new threats to already damaged ecosystems. International attention to this little-known region of Europe is essential to address these menacing developments.

For most of their history, the Evenk preserved their own senses of time and place as well as cosmologies, but the massive industrial onslaught since the 1970s and 1980s has caused great grief and social impacts among the indigenous communities. The main impact has been the destruction of a traditional land use system and cosmology that was inherent, or a unique cultural heritage for the most part, until industrial development. Central to these beliefs and knowledge are the oral traditions and storytelling in the camps and communities of the Evenk.

Despite the industrial projects, the village of Yiengra and the surrounding taiga camps are considered to be of special value in the context of the Evenk home region. Both the customary laws and belief systems of shamanism as well as the local Evenk languages exist to a certain extent. These surviving traditions include the traditional Evenk calendar.

They had successfully defended their values and daily modes of life throughout the Soviet pressures of homogenization, and this had solidified their skills of living inside and according to their stories. The Evenk of Yiengra are still well known in Russia today for leaning strongly on their ethical and spiritual visions in daily life. These traditional ways of life and knowledge, local languages, and natural landscapes form socio-ecological systems, bound and interconnected ways of existence that can still be found from the boreal of the Eurasian North. For example, the Finnish language comes in many forms. Though the internationally established image of Finland rests today on strong social welfare, success at education, and high-tech from Nokia to Angry Birds, it has another, older story. The Finnish language is one of the non-Indo-European languages belonging to the diverse Finno-Ugric linguistic family together with Karelian and various Sámi languages, as well as Komi, Nenets, Khanty, and Mansy in West Siberia. Respected scholars such as anthropologist Matti Sarmela have identified how Finns have undergone a rapid cultural change, from forest Finns into European society, in a space of decades. This process of modernization has left many people affected by the changes. More importantly, most of the forests and natural marsh-mires have been lost to modern industries, wrecking the roots and the survival of deep knowledge, endemic concepts, and cosmology rooted in the old-growth of Eurasian cultures and the taiga.

> Varpuspöllö käänsi päätään odottaen metsämyyrien esiintuloa lumikäytävistään. Kuulosteli valppaasti, terävä katse nuoskaa tarkaten. Mitään ei jäänyt siltä huomaamatta. Pöllö kuin metsä itse, sen olemus ja ilmentymä.
>
> (*Eurasian pygmy owl was sitting on a branch. It turned his head waiting for the bank voles to emerge from their snow tunnels. It was observing and listening with utmost care, sharp eyes scanning the sleet snow. Nothing happened without it being aware of it. The owl was like the taiga forest itself, its being and manifestation.*)

This owl, just as all other old-growth taiga beings and creatures, teaches us to be aware and on the lookout in these turbulent times. The owl manifests the morals, values, and worldview of the forest Finns. The language of our forests may not have a future tense, but it is very much about being in the present tense. Perhaps this is the most potent reminder of our times—we need to be here, now, guided by the forest and the owl. To act, before it is too late for all of us, as parts of the same interconnected whole.

Traditional ecological knowledge, indigenous knowledge, and local-traditional knowledge have received much attention as a topic for environmental and climate monitoring. Scholars often use the terms *indigenous knowledge* (IK) and *traditional ecological knowledge* (TEK) interchangeably. There is a distinct difference, however, between the two terms. The one possessing traditional ecological knowledge does not have to be necessarily indigenous. Such knowledge has been called the knowledge–practice–belief complex, where there is no separation between nature and culture. For the purposes of the Finnish rural villages, the concept of local-traditional ecological knowledge seems to be most appropriate.

TEK discussions have been framed historically through a North American perspective. Anthropologist Julie Cruikshank argues that in order for us to be able to learn anything from personal narrative related to local knowledge, both have to be appreciated in their totality rather than fragmented into data.

Cruikshank has criticized management-driven studies of TEK and environmental parables for often being unable to tap into the range of human engagements with nature—the diverse beliefs, practices, knowledge, and everyday histories of nature that might expand the crisis-ridden focus of environmental politics. She also makes the important point that such studies often organize data to support single arguments that conform to environmentalists' primary concerns about animals, plants, and ecosystems, ignoring what does not fit:

> What looks similar on the surface often turns out to have different meanings and different aims. Codified as TEK, and engulfed by frameworks of North American management science, local knowledge shifts its shape. Sentient and social spaces are thus transformed to measurable commodities called 'lands' and 'resources.' Indigenous peoples then face double exclusion, initially by colonial processes that expropriate land, and ultimately by neo-colonial discourses that appropriate and reformulate their ideas.[15]

John MacDonald, a long-time scholar of the oral histories of the Inuit of Igloolik in Nunavut, Canada, argues, "Inuit traditional knowledge is characteristically personal, its acquisition and application, in varying degrees, specific to communities, families and individuals."[16] His viewpoint, based on more than thirty years documenting Inuit knowledge in this Igloolik community, provides us with a window to a long-running community monitoring and oral history work that is rarely available.

In *Voices from the Bay*, Miriam McDonald et al.[17] documented observations of the Inuit and Cree in the Hudson Bay ecoregion. Their materials are very interesting regarding interlinkages between and across species. Through oral histories and extended community-based monitoring, researchers found links between seal behavior and changes in the sea currents, impact of hydroelectric development on the sturgeon, and additional connected phenomena. We can define these nuanced trails of connections, impacts, and interlinkages as "subtle signs" or weak signals of interconnectivity that have been missing from the natural sciences–driven assessments of development projects in the Hudson Bay region. McDonald et al. also stress the need for a socio-ecological approach, meaning that the cumulative impacts to the Inuit and Cree need to be assessed when development and degradation of natural ecosystems happen.[18]

For oral history, there is a need to build on the successes that have been achieved by creating an ongoing dialogue with traditional and indigenous communities. Particular attention needs to be paid to oral traditions as a means of sustaining the resilience of localized social-ecological systems (SES). Oral histories are critical for sustaining resilience of local SESs. A key feature of such systems is that they also evolve with the system and are not restricted to the rigid formality of Western science.

The difference between living oral histories and documented TEK is that oral histories are communal exercises—the people themselves provide the oral matrix and transfer of knowledge needed to understand the wider spectrum of multifaceted and multidimensional lived knowledge. This allows oral history to provide an authentic representation of local community epistemologies, as opposed to "anecdotes" which often have been taken out of context in the documented TEK materials. Oral histories by the communities themselves are essentially absent from current climate and environmental change discussions.

Indigenous and traditional societies, while sustaining immense damage in the past two to three hundred years to their knowledge and senses of the world, remain storytelling cultures.[19] Traditional Eurasian societies have been studied endlessly since contact with arriving settler societies. For example, the Scandinavian research methods, flowing from the heritage of Russian and German historical research, tend to value written sources. If it has not been written down, it does not exist.

There is a gap, however, between the spoken and the written in the traditional understandings of the world.[20] At its best, oral history does what it is supposed to do—makes invisible histories visible. Such a new reading of indigenous landscapes is both needed and has been recently advocated in scholarship. Oral history materials are valid through their being spoken—and then documented—*events*, through which the traditional people, a community, or the individual have named and recounted their world and its issues.

These events have also been called *earthviews*. Given the rooted co-learning practices of the indigenous and traditional societies of the Eurasian North, we should approach these voices, events, and communal practices as a knowledge category of its own, in relationship and on the same level as Western or European-derived science practices. Earthviews are crucial in themselves, and there is a need to address and listen to them, as they affect the knowledge base of how locals construct reality in a given space, and therefore offer a more diverse and richer understanding of the situation on the ground.

The people participating in oral history fieldwork in a given community have been often pre-selected, or recommended by the local people themselves. This does not remove local biases or conflicts, but it transforms the oral histories from "mere" documents or anecdotes into *communal* and on some occasions *family* processes. The purpose of such community-based work is to respect indigenous governance and mechanisms of *how* oral history and traditional knowledge are handled and discussed in each and every place—it is always different. And oral history is human history, too—it contains its problems, challenges, misrepresentations, and mistakes, just like any other source of knowledge.

Oral history documentation will not solve all problems associated with the hierarchies of power and knowledge production. However, the presence and increased visibility of oral histories is crucial today, because even if the information it contains may challenge the majority view, at least *we have a view* that challenges the established knowledge structures. Oral history work such as this can only provide a window. Such a window can be criticized, dismissed, or applauded, based on the viewpoint of the reader. But communal oral histories allow the hidden to emerge and challenge the mono-truths of history and reality.

Academic criticism is often directed at the communal interpretation and representation of oral histories. Critics may say that the local people speak fulfilling the images of a "golden past" or "best possible stereotype" of their cultures. It is true that biases in interpretation can occasionally lead scholars astray. For example, the Sámi in the 1800s and early 1900s provided (mis/dis)-informative accounts of their lives and cultures across Sápmi. Whether these events were carefully planned cultural defense mechanisms or just plain humor at meeting the arriving peoples is hard to say. However, what we do know is that the Sámi have been aware of the current situation throughout the times of contact and colonization, and have responded accordingly.

Some Sámi authors, such as Elina Helander-Renvall, have made a convincing argument, based on her fieldwork both in Jokkmokk and in Utsjoki and Kaldoaivi areas in Finland, that the Sámi have their own kind of knowledge as well as time-space apparatus. If this is the case, and Sámi oral history needs to be conveyed using these specific means of social capital production and relationships, which manifest best in Sámi language, then there is a clear case for oral history work. Translation of materials into English diminishes the inherent value of oral history, but, on the other hand, in this new language they become available to a larger readership.

Some Canadian partners of Snowchange, such as the Stó:lo, Haida, Kwakwakwakla, Nuu-chah-nulth, and other First Nations, as well as Inuit and Inuvialuit, have successfully won recognition of their oral histories. They have been recognized as a valid mechanism in land use and occupancy, justice, and other research initiatives. For example, the Supreme Court decision of *Delgamuukw* from 1997 in Canada is a landmark event for the indigenous oral histories as a "valid" system of recounting and providing information. Following the *Delgamuukw* precedent, many First Nations in Canada pursued land claims grounded in this case law, which affirmed oral history as evidence.

MacDonald's work in Igloolik from the 1980s to the 2000s demonstrates the high value of long-standing oral history work grounded and based in the community. For decades, he worked to preserve and document local Inuit elders' and peoples' oral histories, on the wishes of the local Elders' society. Only a small portion of this community-approved oral history material has been published so far. Starlore materials that have been made available speak of the high quality of work when the community is in charge of how and to what extent the oral histories are used properly.[21]

Yet, this gap between the spoken world and the written world remains—globally and in Eurasia. Observations provided by K̲ii7iljuus and Harris, two Haida indigenous scholars from western Canada in North America explore the often "hidden" realms of oral history transfers that come close to some of the Eurasian oral history keepers. The experiences from the Inuit, Nuu-chah-nulth, Haida, and other northern indigenous peoples that have successfully engaged with the "Western" histories lead the way to a more holistic understanding of place, time, and events. Though it will be up to the communities themselves to comment and make public, in their terms, whether similar indigenous-controlled and trained systems of knowledge protection and preservation exist, we can review some of these interfaces with science and society to understand the progress made so far.

Alan McMillan[22] identifies the different truth statements and understandings when he explores the cultural heritage of the British Columbian Nuu-chah-nulth, Ditidaht, and Makah First Nations uses of the land and histories. While he recognizes that there are gaps in how and where history has been made and happened, it is important that, in the post-*Delgamuukw* context, he identifies the oral histories as crucial sources of information. McMillan's research on human occupation and changes to the Nuu-chah-nulth societies across time is an important example of putting the spirit and intent of the *Delgamuukw* decision into practice in science, resulting in new, better, and more respectful engagement and results in research.

Six years after the *Delgamuukw* decision, the Haida First Nation participated in a major multidisciplinary investigation into the past of the Haida Gwaii archipelago off the west coast of Canada, close to Alaska. Aspects of this work are worth mentioning here too. K̲ii7iljuus and Harris identify that while oral history research often analyzes the events as either recreated in each generation or unable to transfer knowledge from thousands of years ago in an unbroken line, in fact, the Haida open their cultural doors to provide us with an astounding revelation.

According to K̲ii7iljuus and Harris, the Haida trained a handful of oral historians from each clan and family to learn all of the oral histories of that people. Or, as they say: "The X̲aayda protect the integrity of the oral histories by allowing only those who are properly trained and have the right to tell their clan stories to do so."[23] Only through understanding those hidden histories that can be made public do we start to comprehend the legacies of power imposed on the local and indigenous peoples of Eurasia and beyond.

An examination of oral histories with Kolesov and the Heiskanens provides a method to determine the different stages of the socio-ecological systems in which my co-researchers Kolesov and Heiskanen are located. These systems reveal where the environment and humans are interconnected and are dependent on each other. In these oral histories, their narratives describe stages depicting both intact forests and reflections on heavily modified or altered ecosystems that they have witnessed.

For Vladimir Kolesov, an Evenk hunter and reindeer herder, the surrounding taiga forest ecosystems had been relatively undisturbed from industrial impacts until the 1960s. Naturally, the Evenk had encountered and undergone a range of cultural encounters and changes from the arrival of the Sakha-Yakut in early historical times, to the emerging presence, taxation, and administration of the Russian czar and imperial state, through to collectivization and the Soviet era.

Despite the many changes, the endemic cosmology of the Evenk survived in the forest camps and nomadic herding routes because the Soviets, for the most part, ignored the private rituals and ceremonies, which went underground. Because of this, the Evenk internal rules, customs, and governance of life in the forest was relatively intact in those decades between 1970s and 1990s when Kolesov emerged into adulthood and started to work as a herder and hunter.[24]

As we consider the excerpts from his oral histories, we can see that the Evenk follow their own customary ways of behavior in the forest. This contains mechanisms for interaction in the reindeer camps as well as various features of their landscapes such as mountains, rivers, and so on. Central to the notion of customary behavior are self-imposed limits of harvest—for example, you should only fish as much as you need, not more. The oral history materials are consistent with written accounts of the customary culture of the Evenk that can be found, for example, in the works of folklorist Galina "Keptuke" Varlamova.

The tumultuous era between the 1970s and 2010s, with the various industrial megaprojects close to Yiengra and the surrounding taiga, have left Kolesov to ponder about the future. This cultural act of "narrating the future" is also known among the Evenk, a related reindeer-herding society of eastern Siberia. Kolesov was keenly aware of the impact of the first round of industrial activities that had resulted in a number of suicides and other social impacts in the community in the 1970s.

The renewed activities to harvest the natural resources of the region were, according to him, fatal to the fragile Evenk society that required fairly intact ecosystems of the taiga to function. The environmental change was augmented by the intertwined cultural change of the post-Soviet Russian context. The Evenk language, key to the endemic ways of being with the tundra, was under threat. Combined with this, their religion, or spiritual system, based on their close proximity and relations with the forest, was also dying out.

Kolesov was a member of a well-known Evenk family of epic singers and story narrators, oral historians. For example, his uncle Nikolai Germoginovitch Trafimov was a well-respected singer to whom Kolesov looked up. Alcoholism, loss of economic opportunities, and deteriorating socioeconomic conditions in the remote villages of the taiga combined with the industries and are reflected in the future narration that Kolesov provides.

Across several time zones but still within the same uninterrupted stretch of taiga forest, oral histories of Kauko Heiskanen form a central narrative of ecological change in Finland. For the eastern Finnish cultural areas, Savo and Karelia, winter fishing and hunting activities have been crucial subsistence practices. The oldest archaeological net finding in the world is from Karelia. It is over 10,000 years old. The homelands of the Finno-Ugrian peoples, such as Estonia, Finland, and Karelia are the westernmost outposts of a cultural group that begins in central Siberia and extends from there to the Baltic.

Today, the people in North Karelia identify themselves as "North Karelians" [regionally] or "Savo-Karelians" [nationally] to make the comparison with the ethnicities in the Republic of Karelia, Russia, where the Karelian language is still spoken. The traditional cultures of these Finno-Ugric peoples have much in common in terms of their traditions, languages, and subsistence practices, even though the Sámi are today the internationally and constitutionally recognized indigenous people of Finland.

In general, it can be said that Karelia has always played a significant role in the traditional knowledge and stories of Finnish and Karelian peoples. The Finnish epic *Kalevala* was composed by Elias Lönnrot from the poems that he collected among the community people in White Sea and Finnish Karelia. Many of the songs that epic singers performed for Lönnrot in the Finnish side of Karelia became definitive versions for his book *Kalevala* and thus were, even though changed by Lönnrot, the building blocks of the Finnish modern nation-state.

Therefore, the characteristics of local ecological knowledge building on traditions in the eastern Finnish communities in North Karelia is a matrix of discontinued and ongoing processes. It is on a course of negotiation with the flows from global information networks, mass media, and schooling as they interact and juxtapose with experience and oral history–based knowledge.

The most significant threats to traditions, and the knowledge embedded within them, are the twin forces of rapid habitat change and ecosystem loss[25] since the late 1800s and, connected with this process, a top-down, often forced general knowledge production.[26] This has resulted in the loss of traditional songs and disruption in the endemic worldview and cosmology in North Karelia—both are crucial components of a well-functioning human society.

This erosion of endemic knowledge produces "empty" landscapes ready to be utilized for industrial use, nature protection, or other modern uses. Those communities that still possess fragments of oral histories and relationships with the multifaceted and multidimensional reality of the forest see their lived icescapes as story-filled place-times full of significance and meaning.

Therefore, the surviving local ecological knowledge of the region also maintains the human land use histories, place names of ecological significance and chance, and, in some cases, deep, experience-based ways of being with nature that are themselves under threat from the modern practices. We have to be cautious to take the time to position individual oral histories into the social matrix of a given region and not to assume that all narratives on traditions and sustainability automatically transfer into credible practices on land. Oral history research should not be used to justify all human actions toward nature unless they are "verified," for lack of a better term, to be truly reflective of the sustainability principle and awareness of natural ecosystems. Therefore, fieldwork and collaborative research is always needed to position oral histories in their specific contexts and surrounding situations.

Kauko Heiskanen reflects on the importance of maintaining both lifeways and, in his case, the native cultivated species such as the potato onion and swidden/Swede turnip. His actions have kept the genetic legacy of these plants alive. By engaging in a number of subsistence and commercial activities of the taiga and the Finnish lakes, such as the fisheries of vendace[27] and common bream through the seasonal cycles of the northern year, he has been able to maintain and, in some cases, advance the traditional knowledge of the region.

Stimulus for documenting Kauko's oral histories emerged from the pressure to harvest uranium in his home village. The Kuusoja area contains the richest ore deposits for this mineral used in the production of both peaceful atomic energy as well as nuclear weapons. Heiskanen is aware of the past industrial activities that have permanently altered the local environment, such as the arrival of large-scale hydroelectric stations on the River Pielisjoki. This river used to be the home stream of the unique land-locked salmon, a relic of the Ice Age.

Heiskanen recalls the native fish traps that were used in his childhood to catch these salmon. Now, people cannot access the river due to the hydrostations. In addition to preserving the endemic cultivated species, Heiskanen has initiated, on his own, without any financial or societal support, processes to rewild and restore old-growth forest habitats. This has involved increasing the amount of decaying timber on the forest floor and providing trees of varied age in a forest as opposed to the monoculture forestry practices supported by the timber industries.

In observing the return and nesting of the Eurasian three-toed woodpecker into his rewilded forest, Heiskanen has determined progress, status, and trends in his forest that point to a return to more natural habitat. This can be called a cultural indicator of environmental change. The woodpecker, within the context of Heiskanen's cultural knowledge, represents measures of success in healing the forests from the ravages of industrial logging. Another species referred to in his oral histories is the Siberian flying squirrel. The flying squirrel caused Heiskanen to abandon the industrial logging he was involved in for a time, after the animals had lost their home forests.

K̲ii7iljuus and Harris, for the Haida Nation, and John MacDonald for the Canadian Inuit (more specifically for the Igloolik Inuit) have stressed the communal aspect and control of oral histories. How does this idea manifest in the present materials?

In Yiengra, some of the knowledge holders included, for example, the elder Kim Stepanovich Neystroyev. He only spoke Evenk. Therefore, Vladimir Kolesov was chosen by the Evenk to translate the oral histories of this elder. Similarly, the reindeer-herding brigade 4, the primary partner in the oral history work between 2004 and 2017, chose Kolesov as their spokesperson on the issues of culture, history, and climate change. This points to a role Kolesov had as a person of culture in the community and as a liaison between the elders, the herders, and the outside participants.[28]

Technically, the documentation of the interviews and oral histories were carried out in Evenk, Russian, and occasionally in Sakha languages. Other methods of oral history work included recording of observations and knowledge on maps, field visits to the forest, recordings in video, organization of workshops, and a large-scale international conference, "Snowchange 2007," in Yiengra.

Community workshops were held on the review and use of these oral histories. The focus in the community had shifted between 2007 and 2010, from weather change toward concerns with mining and different industrial land uses, which is reflected in the newer oral histories. All participating holders of these oral histories received their materials back on a CD and may have a future in a community database of oral histories.

For the oral histories of Kauko Heiskanen, the community who initially suggested working with him was the steering committee of the Vaara-Karelia Cultural Heritage Project in Finland. This was a modern community consisting of environmentalists, researchers, fishermen, and concerned citizens. The need to document land use, oral histories, and baselines of environmental change emerged from the renewed uranium mining pressure between 2011 and 2013 in the region. The involved people identified a number of key village representatives to convey traditional knowledge and culture through oral history work.

While Kauko Heiskanen is also known in the wider society for the work he does for the environment, the choice of oral history collection was not directly initiated by the village where he lives. The communal aspect emerged more through the environmentalists and stakeholders of the anti-mining coalition and Snowchange. In that sense, there is a marked difference between the "cohesive" communal decision to choose keepers of oral histories by the communities, as has been the case with K̲ii7iljuus and Harris as well as John MacDonald with the Inuit. This can be partially explained also through the cultural change and the contemporary situation in Finland. Atomized modernity has altered the communal decision-making processes in North Karelia, and while they might have a role to play, with Kauko the decision was more direct. Nevertheless, he continues to be very glad and satisfied with the co-production of written forms of his oral histories.[29]

As the northern areas emerge more into focus due to climate change impacts, natural resource extraction, and geopolitical machinations of various nation states in the twenty-first century, the region's traditional communities and cultures continue to be discussed. Traditional knowledge and oral histories of environmental and climatic change in the Arctic have proven their value in monitoring. For example, the notion of community-based monitoring of environmental change (CBM) and citizen science are some of the emerging trends in partnerships with the indigenous societies of the north.

The words and the fate of Vladimir Kolesov remind us how fragile these northern communities are. As he said, the people of the extreme north have their own laws of living in and off the surrounding nature. These laws need to be respected and followed for the whole system, including both its physical and spiritual parts, to function. Changes imposed through modernization, such as industrialization and its cultural by-products like the rapid rise of alcohol consumption, have dire consequences for these communities.

Beyond the rhetoric of the importance of traditions, livelihoods, and community observations, real cases, characteristics, and dynamics of local-traditional knowledge monitoring, combined with ecological research, are still emerging in the boreal. These oral histories have a chance to assist in more informed views of socio-ecological systems. They are much in line with similar work that has taken place in the North American boreal and Arctic.[30]

The relationship between oral histories of the traditional communities with natural sciences is volatile. In the Arctic context, there have been a number of governmental initiatives to "integrate" findings from "both sides." These steps have included the Arctic Climate Impact Assessment, which contained oral histories of the Sámi from Finland and Russia, to keep the focus on the Eurasian North.

This work was then expanded in the subsequent Arctic Biodiversity Assessment and the Arctic Resilience Report. In the Arctic Resilience Report, a new notion of culturally appropriate indicators of environmental change emerged through the work conducted by the indigenous Skolt Sámi. As in our discussion, cultural indicators such as the work of Kauko Heiskanen, with the rewilding and the woodpecker, would be representative of this.

In drawing the context of living oral histories of these communities into a larger tapestry of nature-based socio-ecological systems, a new issue emerges. If the messages from the traditional knowledge and such histories are to be taken seriously, we also need to guarantee the preservation of the natural environments on which they are dependent. This opens the door for discussions of new governance of lands, waters, and territories of the Eurasian boreal. In the case of materials from Finland and the work of Kauko

Heiskanen, the question of rewilding the "lost" forests back into old-growth habitats may be required as well.

Certainly, this conversation has been going on in the northern areas already for a number of decades. Advent of relational and dynamic approaches of managing these social-ecological systems has created more space for all knowledge, including traditional knowledge, to be recognized as embedded in social and cultural institutions and practices that enable more sustainable resource management. This has been echoed in sustainability approaches, where we see a growing appreciation that co-construction processes are not isolated experiences of knowledge exchange, but are embedded within institutional and societal dynamics.

Particularly in progressive and wealthier national contexts such as Canada, the United States, New Zealand, and Australia, co-construction has now been codified within processes that espouse co-management, described as "the sharing of power and responsibility between government and local resource users."[31] These approaches create a window of opportunity for co-construction to become a vehicle for sharing power with marginalized groups.

In practice, however, the success of co-management is still measured by the extent to which there is formal recognition of such knowledge and oral histories, with little analysis of how this recognition affects broader community processes of self-determination. Anthropologist Paul Nadasdy,[32] for example, shows how formal recognition of the indigenous knowledge of the Kluane First Nation in Canada in a shared governance arrangement led to their knowledge becoming subject to a bureaucratic process based on government-set measures. As a result, a range of external governance mechanisms replaced traditional practices on the land (such as hunting) that were necessary to reproduce knowledge. The formalization of traditional knowledge for the purpose of joint governance, therefore, can potentially lead to co option and assimilation, putting at risk cultural and social practices including embedded leadership and engagement with the land, which are central to the creation and use of indigenous knowledge.

Co-management approaches, therefore, may have moved conceptually toward understanding social dimensions, but they still struggle to overcome the inherent inequality between researchers and communities. Oral histories have a powerful potential to bridge this gap.

In conclusion, the destruction of natural diversity means that the cultural diversity and traditions of a place become threatened as well. Modernity is a mixed bag and needs to be (self-)reflected against the latest discoveries, some of which may take the form of communal oral histories as seen in these two esteemed culture keepers from Northern Eurasia.

Oral histories complement, and sometimes also challenge, the dominant narratives of time, place, history, and reality. They offer a powerful counter-discourse on the expected and "easy" readings of taiga forests and traditional communities. Instead of natural resource areas, these places emerge as homelands with living communities, ways of life and "mind" developed over millennia, based on coexistence and reciprocity with the surrounding nature.

Although Kolesov and Heiskanen have lived in different parts of the Eurasian taiga and utilized different survival strategies, ranging from the nomadic reindeer herding of Siberia to burbot fisheries in Finland, their oral histories illustrate the power and potential of capturing endemic world views. They emerge as powerful counter-narratives to the present discourses of development, environmental destruction, and "empty territories" of the boreal inherent in the plans of natural resource actors.

Evenk of Siberia and the Finns of North Karelia have tried to maintain their specific, endemic concepts and ideas of a life with the taiga, despite the monumental industrial assault underway in both places. In the case of Kauko Heiskanen and his oral histories, a step further has been taken—he is narrating and shaping already the restorative actions through the rewilding of forests and re-emergence of the Eurasian taiga as a healthy home of all living beings. Therefore, the work with oral histories benefits the place, the people, and the environment.

For the century underway, preservation and restoration of the traditional mind of the taiga remains the greatest challenge. Success remains to be seen.

Notes

1. The Taiga boreal ecosystem is a vast expanse of forest from the Pacific Ocean in Siberia to Finland in the northwest corner of the continent. While being impacted by forestry, industrial actions, and other changes by humans, large tracts of the boreal constitute a major biodiversity hotspot and a climate sink.
2. The Snowchange Oral History Archives are located at the Snowchange HQ in North Karelia, Finland. They are community-controlled, i.e., the specific person, family, and community, depending on the context, has reviewed and shares copyrights to the materials and access to them. Requests should be made to contact@snowchange.org for cooperation and access.
3. More precisely, the Evenk scholars Galina Varlamova, Anna Myreeva, and Tamara Andreeva had chosen Yiengra as the location for new oral history work due to its rich

cultural heritage, preserved nomadic lifestyle, and language. The three researchers had conducted internal oral history work there before and collaborated with international and Russian scholars. Once the research team from Snowchange and the leading Evenk scholar, such as Tamara Andreeva, had been working in the community for some years, the people of the village pointed certain individuals to participate in the partnership.

4. *Obshcina,* clan enterprise/tribal community, and other forms of social self-governance are forms of self-organization of individuals, belonging to the recognized Russian Northern Indigenous peoples, united on the basis of blood kin relations (family, kin) and/or territorial-neighborhood principles, created with the aim to protect their original territories of inhabitance and to protect and develop their traditional lifeways, economies, and culture.
5. While there are many legislative pieces regarding the *obschiny*, we can investigate their English-language definition using the Federal Law No 82-F3 1999, which defines the *obschina* in the following way: Obshchiny and other forms of social self-governance are forms of self-organization of individuals, belonging to the numerically small peoples, united on the basis of blood kin relations (family, kin) and/or territorial-neighbourhood principles, created with the goal of protecting their original home territories and to protect and develop their traditional ways of life, economies and culture (Federal Law No 82-F3 1999, Russian Federation).
6. Snowchange Yiengra Oral History Tape 2005–06.
7. See more on narrating the future from the linguistic neighbors, the Even, in Olga Ulturgasheva, *Narrating the Future in Siberia: Childhood, Adolescence and Autobiography among the Eveny* (New York: Berghahn Books, 2012).
8. Snowchange Yiengra Oral History Tape 2005–06.
9. Aggregatum Group of *Allium cepa*, known in Scotland as "Swede Turnip."
10. *Brassica rapa.*
11. Snowchange Eno Oral History Tape 2013. All of the accompanying materials are from Kauko's oral history materials.
12. *Pteromys volans.*
13. *Picoides tridactylus.*
14. Available at http://www.lumi.fi/vaara-karjalan-kulttuuriperinto/ online.
15. Julie Cruikshank, *Do Glaciers Listen – Local Knowledge, Colonial Encounters & Social Imagination* (Vancouver: UBC Press, 2005), 259.
16. John MacDonald, *The Arctic Sky: Inuit Astronomy, Star Lore, and Legend* (Toronto: Royal Ontario Museum, 1998), 6.
17. In Miriam McDonald, Lucassie Arragutainaq, and Zack Novalinga, *Voices from the Bay: Traditional Knowledge of Inuit and Cree in the Hudson Bay Ecoregion* (Ottawa: Canadian Arctic Resources Committee, 1997).
18. In other regions, the concept of "citizen science" has emerged as a related topic. Central to citizen science (CS) is the idea of residents of an ecosystem or a place conducting research and monitoring regarding a given topic. The partnering natural sciences need, and initiative often drives, CS efforts. However, for the purposes of this chapter the local-traditional knowledge concept with its focus on the oral histories

and long-term engagement with the local places is used. They retain the agency of the local community, in contrast with CS.

19. As Chinua Achebe, the celebrated Nigerian author, provides in his *Things Fall Apart*, transition from oral, spoken systems of interaction with the world into written forms is often a violent, imposed process. Chinua Achebe, *Things Fall Apart* (New York: Anchor Books, 1994).
20. For example, it has been often said among the Skolt Sámi that their own culture, worldview, and ways of being with the world do not get proper attention, because their oral histories are not made visible.
21. Starlore relates to knowledge of celestial events such as stars, moon, northern lights, halo, and so on.
22. Alan D. McMillan, *Since the Time of the Transformers: The Ancient Heritage of Nuu-chah-nulth, Ditidaht and Makah* (Vancouver: UBC Press, 2000).
23. In Kii7iljuus (Barbara J. Wilson) and Heather Harris, "Tllsda Xaaydas K'aaygang. nga: Long, Long Ago Haida Ancient Stories," in *Haida Gwaii – Human History and Environment from the Time of the Loon to the Time of the Iron People*, ed. Daryl W. Fedje and Rolf W. Mathewes (Vancouver: UBC Press, 2005): 123.
24. He also received formal education in an institute in Northern Yakutia in the Soviet period in subjects of reindeer and animal husbandry.
25. Loss destruction of old-growth forests, hydroelectric power, impacts to water ecology from peat production, and pulp factories.
26. Here referring to the state-led schooling, especially in those regions of Finland where the oral history and traditional governance structures still played a major role in the 1800s and early 1900s.
27. A salmon-related schooling fish, *Coregonus albula*. A highly regarded local traditional food.
28. In recent years, the life of the Brigade 4 has been documented in a German film for television. Available at https://www.youtube.com/watch?v=zC_c8ZUBz04.
29. Timeline for the documentation was rather short—all was done during 2013. In addition to the digital recording of the oral histories, we conducted joint field visits to his forests, ate the potato onions and swidden turnips, and visited several times. The products of the oral history work included a workshop, exhibition of posters and booklet, as well as media coverage. Translation of the Finnish interviews into English took place thanks to Snowchange and was reviewed by native English speakers. By 2018 the immediate plans for a uranium mine had been withdrawn for the time being.
30. See more in John MacDonald, *The Arctic Sky*. He demonstrates the value of community-based oral history project to add to the complexity and depth of oral histories as a mechanism of observing change.
31. Fikret Berkes, *Sacred Ecology: Traditional Ecological Knowledge and Resource Management*. (London: Routledge, 1999).
32. Paul Nadasdy, *Hunters and Bureaucrats: Power, Knowledge and Aboriginal-State Relations in the Southwest Yukon* (Vancouver: UBC Press, 2003).

12

Who Speaks for the Trees?

Forestry in the Scottish Highlands

K. Jan Oosthoek

Oral histories collected by Jan Oosthoek with retired foresters of the British Forestry Commission, Stirling and Edinburgh, Scotland, June 1998–August 1999

Anyone who travels through the Scottish Highlands is struck by their bleak beauty, empty moors, and rugged mountains, but also by the large blocks of single-species (monoculture) conifer plantations.[1] *The creation of these forest plantations is probably the biggest change in land use since the agricultural improvements of the eighteenth century, the so-called Highland Clearances. The British Forestry Commission created most of these plantations from its inception in 1919. In recent decades, environmentalists have decried the blocks of monotonous conifer plantations on the Scottish hillsides and criticized the foresters that created them for not having regard for landscape aesthetics and nature conservation.*

In 1998 and 1999, I interviewed some of the men who created these forestry plantations as part of a PhD project studying the history of the creation of new forests on the bare lands[2] *of the Scottish Highland. The aim of the interviews was to find out how these foresters had created the new forests and what they thought of the criticism leveled at them by environmentalists.*

Interview with Jim Atterson, University of Stirling, Scotland, August 13, 1999

Jim Atterson started his career in the Research Branch of the Forestry Commission in Edinburgh. He worked as a district officer in the Dornoch District in northern Scotland before he took up the position of Principal Silviculturist North in the early 1970s. At the beginning of the 1980s, he

K. Jan Oosthoek, *Who Speaks for the Trees?* In: *Oral History and the Environment.* Edited by Stephen M. Sloan and Mark Cave, Oxford University Press. © Oxford University Press 2022. DOI: 10.1093/oso/9780190684969.003.0013

became assistant conservator, harvesting and marketing for the West Scotland Conservancy before becoming the Conservator for Scotland. He retired in the early 1990s.

OOSTHOEK: The general public started to object to conifer plantations in the 1960s. What is your recollection of this development?

ATTERSON: I don't remember that it was as early as that while working for the Forestry Commission. I started in 1959. Right through the 1960s and 1970s I don't remember much being said against planting in Scotland by the general public, even by conservationists because there were not that many conservationists in those days. Frank Fraser Darling[3] did certainly not object [. . .] to forestry [and] he was even saying that some of the conservation groups were not correct in complaining about what foresters were doing. I get the feeling that only in the late 1970s and then really in the 80s, it really started, and on the deep peats it did not start with the Flow Country.[4] I took over the North Conservancy in 1985, and shortly after that we were asked to stop ploughing deep peats. Just overnight. That took us by surprise because we had been planting these very big programs from the mid-1960s onwards on deep peat. When I was in the Dornoch District in the late 1960s and early 1970s we were planting over three thousand acres per annum. Most of it was deep peat and no one said anything about it. We thought we were doing a great job because were creating a forest where there had been very degraded peat lands because of periodic burning. The ground was used for sheep, deer, and grouse depending on what the estate was interested in and the economic interest at the time.

OOSTHOEK: Strathyre in Central Scotland was planted from the 1950s onwards and looks beautiful. Wasn't it landscaped at the same time?

ATTERSON: The forester at that time, [Don MacCaskill],[5] planted a variety of species according to site type. He wasn't specifically landscaping, but he was planting different species on different sites and because site types tended to follow the topography, different soils in valleys running up the hills, you got different species.

OOSTHOEK: That is according to Mark Anderson's method of matching tree species to the site?

ATTERSON: Mark Anderson[6] wrote a book called tree selection, and that was based on planting trees according to soil type. That is what the forester at Strathyre was doing and a lot of other foresters as well. But the further north you went, up to the poorer soils, you could not really do that, because you were very limited to the number of species you could use: pine and spruce. That idea still continues in the north: you tended to plant pine on the drier ground and spruce on the wetter ground. Until you get to the far north where even spruce would not grow and you had to plant almost pure lodgepole pine (*Pinus contorta*). The idea was that lodgepole pine is the only tree that grows up here, which can send its roots down into anaerobe peat. It does this by creating air passages in its roots so it takes oxygen down its roots. If you dig up a piece of lodgepole pine root, and chop off a piece, you can actually blow bubbles through into water, like a straw. Lodgepole dries the peat out and a lot of these peats are called a pseudo-fiber peat that means that they look a bit fibrous. But when they dry they change and you rework them. This enables the peat to be burned in fires. Lodgepole pine does this with the peat in situ and when the peat dries it shrinks. Cracks are being produced in the peat and this means that the water can get away much quicker. The benefit of all this is that you can then plant other species.

OOSTHOEK: Were you involved in forestry economics?

ATTERSON: Only as a manager but not with the policy part of economics.

OOSTHOEK: What I noticed is that forestry economics became important during the 1960s. Why?

ATTERSON: Well, they appointed a forestry economist at the time. Hiley published a book on forestry economics. That was very early on [in the 1930s].[7] At time the government wanted to know what they could expect in return from the forests with regard to the money invested.

OOSTHOEK: Didn't you think that the introduction of this kind of economics prevented the Forestry Commission from creating a more diverse forest?

ATTERSON: That is right, because cash had to be accounted for and there was only a small percentage which was put aside for things like landscaping and conservation. The bulk of it was for creating forests to produce money at the end of the day. That was economics driving forestry. Foresters found it very difficult to build in a lot of what they would like to build in the way of variety of species, bare land, land planted for one reason or another because the maximum return had to be got from the forest during that period, or the maximum expected return.

OOSTHOEK: Isn't that in fact violating how ecosystems function? You can dictate things to a certain level.

ATTERSON: If you're thinking of the whole ecosystem and wanting to use the ecosystem as varied as possible and to have the maximum amount of life in it, life of all kinds, that is true. But at the same time the people spending public money, wanted the maximum return on it. It is the same as a farmer. He is not planting a field of wheat to the benefit of the ecosystem but to get the maximum return in cash.

OOSTHOEK: But in forestry there are two sides: making a profit and the ecological value.

ATTERSON: That is right. It is very difficult [to determine] what the [ecological] value is to the country because there is a monetary value to the owner, there could be intrinsic benefits. But the actual creation of a diverse ecosystem has no value to the actual owner but it probably does have to the country, so the country should pay for it, not the owner. That is partly true to the grants that were available to landowners to create a multi-purpose forest.

Before the present day there wasn't the public perception that this was a good thing. The public perception at that time was the governments' idea of public perception: the maximum economic return. In the long run, and that is what Frank Fraser Darling said, you get varied forests because of all kinds of things, like fire, wind and so on, creating an irregular forest. Then you get a whole range of species regenerating naturally.

OOSTHOEK: What was Frank Fraser Darling's view on forestry economics? It does not show in his work.

ATTERSON: No, that is not the kind of thing he had a view on because he was not interested in it.

OOSTHOEK: That is contrary to Mark Anderson who saw both economics and ecology in forestry. Are these two areas not conflicting?

ATTERSON: No, not really. I think that Mark Anderson . . . his view was that a multi-purpose forest was a forest that could withstand problems such as wind and disease, better than a monoculture. So, he was looking at it as a forester and realized that a multi-purpose forest was in fact a better forest in producing, in the long run, timber. Incidentally it also produces wildlife, but I don't think that was his intention; it was a bonus.

OOSTHOEK: Someone is not going into forestry to just grow trees. They must be interested in other aspects such as wildlife.

ATTERSON: Well exactly. I got into forestry because I was interested in the whole range of what is now called conservation. We called it natural history.

OOSTHOEK: Can you say that foresters were pioneers?

ATTERSON: Very much.

OOSTHOEK: But weren't foresters overtaken by events?

ATTERSON: They were overtaken only in that they did not have enough money to do the things that would have to be done to create a multipurpose forest quickly. That's by doing premature felling, planting a range of species which wouldn't be as economic as the species they planted, and some would not have been economic at all. There was no money to do this in the early days.

OOSTHOEK: Was that because the objective of forestry was just to create a strategic timber reserve and that's it?

ATTERSON: Yes, but even when the strategic objective disappeared the idea was still to create a forest that had maximum economic value.

OOSTHOEK: During my research, I noticed that foresters at lower levels started to create forest walks for the public, although there was no money for it. It was not economic.

ATTERSON: Well, when you said it was not economic, it was done when men were available when there was no other job to do. They could use that time to do that job, such as preparing forest walks and viewpoints, and so on. Foresters did plant amenity species in the forests. Foresters have always been keen on planting a variety of species because it makes the job more interesting. What was holding them back was just a lack of funds.

OOSTHOEK: My impression is that initiatives to create facilities for the public came from the bottom and not the top. It was not official policy. It only became official policy when public demand for access increased.

ATTERSON: It was just done. It's just that there wasn't a policy from the headquarters or conservancy offices that this will be done in specific areas.

OOSTHOEK: But at a certain point in the 1960s recreation is popping up as a policy issue.

ATTERSON: Yes, that is right. As public opinion changes, the government's view of what should be done changes and more cash is available to do things that the government wants to be done. The government does things what the public wants done. Prior to that you could not spend a lot of cash preparing forest walks, viewpoints, etcetera. But once the government had the view that the public wanted this, there was some money made available to it and it was a gradual increase in money annually.

OOSTHOEK: Here we have come to the end of the interview. Is there anything you would like to add?

ATTERSON: The first time we talked on the phone, you gave me the impression that you thought that this idea of the foresters at the sharp end having the idea that there should be more recreational and conservational etcetera before headquarters did. I don't think that is correct. The foresters at headquarters had exactly the same ideas, and it is just that the foresters on the ground could actually do something about it, although be it in a very small way. But they could not really do it as part of their main program, because it wasn't in the main program. It wasn't budgeted for, so they had to do it as an extra.

OOSTHOEK: Bob Allison[8] told me that one day the conservator of Aberdeen, Mr. Oliver, suggested to fence forests to keep people out and that Bob reacted by saying that there was no reason to fence people out. He argued that it was better to educate them than to keep them out because that's cheaper. He said that was a typical reaction. He also said that it was cheaper to use trappers to exterminate rabbits instead of raising fences against them, and that is what he did. He got advice from the conservator to erect fences. About the winding walks he commented that in fact it was not encouraged and that it was initially criticized because he was putting money in it.

ATTERSON: You are talking about personalities there. I knew both of them in East Scotland at that time. . ., probably discouraged the public from going into the forests more than anywhere else in Scotland. But that was because of personalities, not because of an overall policy as such. I don't think you should read very much into that. That certainly did not occur in the other three conservancies. I knew this because I was in research, covering the whole of Scotland. So, I could see the differences between the conservancies.

OOSTHOEK: Because all the people who worked in the East talk about these things: recreational facilities, etcetera, were not encouraged.

ATTERSON: That is because of one person. And over a relative short period. You can get these differences between general managers, if in fact you call them that; you know, you bring in a new manager and policy changes quite considerably. So, that is what was happening there, so don't read too much in that because it was just one person's opinion of what should be done. It certainly was not Forestry Commission policy.

OOSTHOEK: How was it then in the other three conservancies?

ATTERSON: Public access to the forest was free and easy. To give you an example: in Research, we had to get permission from Aberdeen to go into any of these east of Scotland forests. It did not happen anywhere else.

Interview with Fred J. and Margaret Ann Donald, Edinburgh, Scotland, July 13, 1999

Fred Donald graduated in forestry from the University of Aberdeen in 1950 and was then employed as a forest surveyor. He worked from 1951 until 1957 as a private woodlands officer in Banff before appointment as a district officer of the Forestry Commission, a position he held in several places before he retired in 1980. Fred married Margaret in 1957.

OOSTHOEK: Economics became more important after the Second World War because of the fact that more trees reached maturity and were ready to be harvested. A recent study suggests that forest economics were very single-minded.[9]

F. DONALD: I think, basically during the early 1980s we did not understand forest economics, because the whole forestry business had been taken over by events. The forests had been felled because we needed the timber, irrespective if it were a good or a bad thing. We had to put them back and people were not saying Sitka spruce (*Picea sitchensis*) is more economical than Scots pine in these days, they were just saying, "this land is good for Scots pine and this land is good for spruce or this land is good for larch." Larch [seemed] a highly profitable crop before the war, [but] it became evident that it was not particularly profitable because it occupied too much ground for the amount of timber it produced. This was rather a gradual change of thinking on these lines. Artificial prerogatives took over from . . . woods that would be used in coalmines.[10]

OOSTHOEK: Now about the relationship between headquarters (HQ) and people working on the ground in the forests. Graham Tuley[11] remarked on this that, "the districts were a kind of kingdoms." He meant, the person who were in charge of these forests decided what happened. Orders came down from HQ and local foresters were interpreting them on their own terms. They said, "It's very nice but we do it this way." Every district officer did things his own way and in doing so produced a large diversity. It that impression correct?

F. DONALD: Yes, up to a point. You had the assistant conservators who came around and asked, "Why are you doing it this way?" By large they controlled the plant supply. The district officer had to consider what land we were going to plant during the next five years. We made forecasts of plant requirements relative to that. They organized what was to be sown in the nurseries and the plant transplants and so on. We never knew what was available. Eventually we were told, "You take Douglas fir, or you will take lodgepole pine or larch or what they had a surplus of." You had to ask, "What am I going to do with this stuff?"

Well in many cases, that was it. If you go to Spey Mouth [in northern Scotland], you will find the unusual mixture of lodgepole and larch because we did not have much ground to plant larch on, so I put the lodgepole pine and larch together.

OOSTHOEK: This reminds me to a section in J.A.B. Macdonald's[12] unpublished memoirs mentioning that many trees died because they were not planting the right trees on the right soils.

F. DONALD: Oh, there was a lot of that. [Macdonald] was experimenting with suppressing heather, among other things, getting spruce to grow by growing a mixture with pine, or cultivating getting spruce to grow where we would have difficulty growing pine. So, we developed our culture of mixing pine and spruce in the northeast of Scotland. Around 1970 research said "this is not on," but after I retired they resurrected the idea. This was particularly true for a non-thin situation. If you planted an area with spruce and they all lived, you had a thinning problem. If you planted Sitka spruce and pine and the Sitka spruce outgrew the pine, the pine would [initially] suppress the heather and [after that] you would have a first thinning without doing anything. This is a cheap way of thinning and . . . for some reason or another the Research branch could not digest [it]. They saw it as [the trees] all growing up together and you were manuring all these trees and it was a bad thing to manure the trees that were going to die anyway. But, a proportion of the spruce were going to die anyway if you planted pure spruce. It was a cheaper way in some ways of establishing spruce on fertile heather sites than trying keep the heather suppressed all the time, because if you planted pure spruce on a heather site not only you had to burn the heather off before you started, you had to go back and kill the heather at intervals until your spruce was established. There was this shuffling of ideas. At one time it was a good idea and at another a bad idea. So, I was stopped from planting mixtures.

OOSTHOEK: During the last twenty years the Forestry Commission has been criticized for growing monotonous forests of mainly spruces. At the time that the Forestry Commission was established in 1919, the only land available for forestry was poorer and elevated sites. Spruces were the best trees to grow there. If you wanted to plant something else it was simply not economic. Is this interpretation correct?

F. DONALD: If you did basic net discounting revenue calculations, Sitka spruce obviously came way ahead of Norway spruce for instance, and streets ahead of larch and pine. So, the temptation was to push spruce as far as it would go.

OOSTHOEK: It was an answer to the policy objectives?

F. DONALD: It was producing revenue faster and it was producing timber, which was short in supply.

OOSTHOEK: This was all shortly after the Second World War. But when did environmental considerations come on board?

F. DONALD: That came in around the 1970s. We started looking at forest design from a visual point of view, taking on board the advice of Nature Conservancy Council, to diversify the forests somewhat.

OOSTHOEK: At the start of the 1960s, the Forestry Commissions Annual Reports mentioned for the first time environmental considerations. Was it really implemented at that time?

F. DONALD: I think that was for public consumption. It was to show that we were the good guys and not the bad guys. But it did change. There were, within the Forestry Commission, people saying, "We must retain more birch when we are planting trees." Not only it looked nice, but also because it conserved the soil better then Sika and Norway spruce for instance. There was a greater concern for wildlife when we found a pine marten carcass in Spey Mouth, which created a great deal of interest because we hadn't had pine marten in that area in living memory. That sort of thing. There was Donald MacCaskill for instance, he knew a great deal about wildlife. He was one of my foresters for a bit and he has published books on the subject and his photographs are very well taken. There was a groundswell of it somewhere, although it wasn't necessary official policy. It came from the fact that people were going into the forests in the first place. People love the forests with everything in it.

We should get rid of the idea that we cover Scotland with trees at the lowest possible price. You have this other thing, why were the men who were planting these forests doing it? Not just for the money but for something to create.

M. DONALD: Can I just say Fred, he [MacCaskill] was an expert on deer, staying up all night watching [them]. This was a forester who did it for his own, staying up all night just to see deer coming out in the early morning in the mist and study them, as MacCaskill did with foxes and ravens and all sorts of things. Obviously, it was there in the people working in the forests. They were the enthusiasts.

OOSTHOEK: But, is this a tradition from earlier because during the 1920s and 1930s they are talking about wildlife and the like? It was said that a forest is not a "timber factory," it is more than that, it is everything that comes with it.

F. DONALD: Going back to the 1940s and 1950s, the objective was to get the forests established and other things were secondary. Because people had gone into forestry in the first place, they wanted to do forestry. Their natural instincts started to develop. Interest in wildlife or vegetation because of the fact they liked the forests.

OOSTHOEK: How did the silvicultural systems you were using develop over time? I think there was a development, a change.

F. DONALD: There was a lot of experimentation going on with, for instance, thinning larch with underplanting it with spruce. It was not a good idea because it was hard to get the larch without damaging the spruce. But to what extent was clear-felling justified? This was becoming a consideration. The idea was that you wouldn't clear-fell fast areas in the wood. This was about early 1970s. A very recent development. Basically, the economics are that you plant this thousand acres within a six-year period and you fell it in eighty years' time. But it creates the same problem that we inherited that you have, from an environmental point of view at least, a forest condition.

OOSTHOEK: At the present, conifers are removed to clear the broadleaves that were once underplanted. That does not sound very economic.

F. DONALD: It is gardening on the site if compared with the bigger picture. It is the Silvia Crowe[13] thing. She was a sensible woman, don't decry Silvia Crowe because she acknowledged the need for forestry to be profitable. She asked, "What is the minimal you could do to make it look a bit better?" She sold this idea quite well. If you doing a forest switch as you are facing across a valley for instance, think of what it would look like from the other side. Too many straight lines, cutting edges . . .

M. DONALD: Or having great public outcries on the road to Skye, obviously great blocks on hillsides, people were inviting the press, they were ugly. And the work of Silvia Crowe was the result, because they were so ugly people objected to them. It was their money that was being spent, taxpayer money. There was political pressure.

OOSTHOEK: I have been looking for evidence for this public outcry, but I only find a few instances.

M. DONALD: Maybe the Commission did not record it, but people did write to the newspapers. I have been reading this in the newspapers, they did write and complain about the ugliness of these great blocks that you could see across a valley. Why did we have to have this sort of thing? We are not talking about some person from some council house somewhere, we are talking about people with knowledge of trees and environment objecting to this sort of thing. Maybe the Forestry Commission never recorded it, but it did happen. In these days, before it became the Forest Enterprise, the Forestry Commission was dependent on public money for doing what it did. It couldn't even prove that what it was doing was profitable. Scottish forests have a reasonable chance to be profitable, but down in the south they were running gardens.

Commentary

At the turn of the twentieth century, less than 5 percent of Scotland was covered with forests, a figure that was not much higher in the rest of Britain. Forest cover in Scotland had not changed much since the Middle Ages, when most of the land was already cleared for agriculture. Since then, woodland cover has remained surprisingly stable due to careful management of the forests using techniques such as coppicing, the periodically cutting back of a tree to ground level to stimulate growth, and strict laws limiting how much wood people could take. When Britain industrialized in the nineteenth century most of the timber needed for construction, mining, and industrial purposes was imported. By the turn of the twentieth century the country imported more than 90 percent of its timber needs, but wood imports did not worry the British government. The United Kingdom was the only country in Europe without a formal forestry policy in 1914.[14] In normal times importing timber from overseas was no problem, but during the First World War, the German naval blockade made import of timber very difficult. As a result, Britain relied on its own limited forest reserves to supply the timber needed in the trenches and the mines to keep the war effort going.[15]

It became clear to the British government that having so little woodland is a liability during times of war. After the war, the government formulated three objectives for a British forest policy: maintain an adequate reserve of standing timber to supply the country in times of war; make better use of uncultivated and derelict land; and foster the general well-being of rural Britain, including rural employment.[16] These three objectives remained the foundations of British forest policy for almost half a century. In 1919 the government created a state forest authority, the Forestry Commission, to supervise the proposed planting program. The efforts of the Forestry Commission were highly successful, and, by the early twenty-first century, tree cover rose to 17 percent of Scotland's land area.[17]

The Forestry Commission planted most of these trees on open, peaty moorland, and sometimes on older birch and oak woodland sites. The main type of trees planted were conifers, originally from North America, because these trees grow better and faster on the upland moorlands than the only native conifer, the Scots pine (*Pinus sylvestris*). This surge in afforestation is probably the single largest change in land use in Scotland since the agricultural improvements of the eighteenth century.[18] The work of the Forestry Commission turned the upland moors into spaces of production through the introduction of conifers and the practice of "scientific forestry." By the second half of the twentieth century, the Highlands of Scotland had become an ecotechnical environment, with the main aim to produce timber as efficiently as possible.[19] The visible result of production forestry was the creation of large monoculture forest plantations blanketing the hillsides. Since the dawn of modern environmentalism in the 1960s, this type of production forestry has come under intense scrutiny. Scottish foresters were often made out as technocrats who could not see the wood for the trees[20] and regarded such values as nature conservation and landscape aesthetics as a threat to the efficient production of timber. There is some truth in the observation that foresters were obsessed with the creation of plantations and not forest ecosystems, especially in the decades following the Second World War.

When war broke out in 1939, most of the Britain's forest plantations were less than twenty years old and not mature enough to supply British industry and the war effort. The war made the import of wood difficult; Britain experienced wood shortages and relied on its own woodlands. After the war, this triggered a strengthening and acceleration of the program of forest expansion. This expansion was very much focused on the creation of large-scale single tree species plantations with the increasing use of mechanical ground preparation and the input of chemicals.[21] More than half of the expansion of forestry took place in Scotland, which is only a third the size of Britain (Scotland,

England, Wales, and Northern Ireland). The transformation of large tracts of the Scottish landscape into geometric forestry plantations is very visible.

By the 1960s, wartime imperatives to expand domestically produced timber weakened, and a more affluent and increasingly environmentally conscious public started to question the methods used in modern forestry. This pressure was initially resisted by the forestry community but, surprisingly, by the 1990s forestry in Scotland took an "ecological turn" with emphasis on the development of healthy eco-systems and conservation.[22] Even then there was a lingering belief among many conservationists that foresters were interested only in monoculture plantations and were damaging the landscape.

Perhaps it is no surprise that in 1997, when I started a PhD project investigating Scottish forestry in the twentieth century, hardly anyone had asked the people who created the new Scottish forests how they did it and what they thought about it. Neither had anyone asked how these foresters facilitated the greening of forestry in Scotland and what they thought about nature conservation and landscape aesthetics. I suspected that it was likely that the swift "ecological turn" of forestry would not have occurred so fast if the ideas of nature conservation and landscape aesthetics had not already been present among the forester corps of the Forestry Commission. To find out, I decided to interview retired Forestry Commission foresters ranging from the very top of the organisation, a former director general, through district officers and foresters working on the ground. These first-hand testimonies document the personal stories of the people that created the forests that are rarely available in the official record.

In the 1920s, in the early days of the Forestry Commission, many foresters viewed the uplands available for forestry as a landscape with little beauty or value. Foresters and landowners believed that tree planting would improve the economic value of the land. As a result, wider issues such as landscape aesthetics, conservation, and even the creation of recreational facilities were not part of official forestry policy. With the wood shortages experienced during the two world wars, it is no surprise that this shortfall preoccupied the minds of the founders and subsequent leaders of the Forestry Commission. The remit of the Forestry Commission was to create a timber reserve as fast, efficient, and cheap as possible. The only problem was that the land available for forestry was not very suitable for growing trees.[23] The land was often of poor quality, exposed, waterlogged, and subject to very low temperatures during the winter. However, over time foresters learned to cultivate and plant these upland areas successfully. They were very proud of these achievements and this came through in the interviews.

Both Jim Atterson and Fred Donald point out that a lot of research and experimentation was going on and that Forestry Commission foresters became very good at planting on a variety of difficult sites by using the right tree species and cultivation techniques. They had a deep knowledge of trees and how to plant and look after them. During their experimentation, it became clear that the best and most efficient way of creating new forests was by cultivating the land using deep plowing, applying fertilizers and planting monocultures, forests composed of a single species. The official policy aiming at the quick creation of a timber reserve was the overriding consideration and that left little room for developing other functions of the forests. This created the impression that foresters were not capable or interested in doing anything beyond growing trees as quickly and efficiently as possible. But, according to Jim Atterson, that makes a mockery of the reasons why people went into forestry, "I went into forestry because I was interested in what is now called conservation." But he justified the fact that not much was done to provide recreational facilities, create a species diverse forest, and encourage conservation because there was simply not enough funding. Atterson observed that foresters found it difficult to do all the things they wanted to do, such as creating forests with varied species, provide recreational facilities, and protect wildlife. Fred Donald highlights recreational and ecological values of forests as well by pointing out that there is more to forestry besides planting trees. He believed that the motivation of the people planting new forests was not just having a job to earn money but "for something to create." This is a profound insight into the justification for the work that these foresters did: to create something for future generations, not just to make money or as a resource but also a natural place of beauty and enjoyment.

The interviews with retired foresters confirmed time and again that they were interested in more than just growing telegraph poles. Fred Donald recalled the excitement when they found a dead pine marten in Spey Mouth, an animal that had not been seen in that area in living memory. Various people also recalled Donald MacCaskill, head forester at Strathyre, who studied and recorded small mammals and birds living in the forests under his care. MacCaskill was a keen naturalist and wildlife photographer who published books on wildlife and birds and helped the BBC with the production of wildlife documentaries.[24] He designed the forests so that they supported wildlife and resisted the removal of old growth woodlands. His superiors did not always approve of such methods. Early in his career, MacCaskill received the order to cut down some old growth oak woods on the banks of Loch Awe but refused to do this because he believed that these trees needed protection. This attitude amounted to a refusal to carry out formal forestry policy, but his persistence won in the end and he saved the oak trees.[25] MacCaskill was probably one of the foresters who most openly challenged some aspects of official protocol, but it was not uncommon. These anecdotes of a dead pine marten and his refusal to remove trees make forest history more vivid and call attention to the fact that foresters viewed forests as living and breathing ecosystems. These distinctive responses in oral testimony bring out the character of the historical actors involved and adds the sounds, sites, and smells associated with these foresters and the landscapes they managed. These associations with landscapes and forests applies to any level of a land management or environmental organization. Jim Atterson suggested that even at Forestry Commission headquarters there were many who believed in developing more diverse forests, with attention to recreation and conservation.

Yet, they were bound to formal forestry policy and could not do much to change this. It was working foresters and their managers on the ground who could make a difference and do interesting things not directly related to timber production, if time and money permitted. As a result, the provision of recreational facilities and encouraging forests that were more diverse was patchy and varied a great deal throughout Scotland and often depended on personalities in a particular forest district or conservancy, as Donald and particularly Atterson pointed out: "You can get these differences between general managers . . . ; you know, you bring in a new manager and policy changes quite considerably."

Such clear interpersonal tensions within the narrative of twentieth-century forestry are difficult to unearth using the official archival record. Oral testimony is able to illuminate such tensions and diversity of voices within the Forestry Commission, or any land management agency. It gives granular detail to the ways in which working practices are contested and change over time within land management organizations, often related to intergenerational change. Oral history also highlights personal preferences by uncovering tensions between outcomes that some foresters might consider desirable while others see as bad practice. The value of oral history is that this kind of tension brings color to our histories and at the same time helps us to avoid simplistic declensionist environmental history narratives of decline and destruction of landscapes and environment.[26] That does not mean we should avoid writing in our histories about the negative consequences of afforestation, but the declensionist aspects should be counterbalanced with narratives of events that come as close as possible to actual lived experiences and interactions with forest ecosystems.[27]

Another contentious issue was public access to state forests. Although many foresters believed that public access to the forests and the provision of recreational facilities was a good thing, it was not until the early 1960s that it became part of formal forest policy. Most foresters interviewed believed that public opinion drove the introduction of this new policy due to an increasingly automobile and environmental conscious public, as Margaret Ann Donald maintained, "having great public outcries on the road to Skye, obviously great blocks on hill sides; people were inviting the press; they were ugly."

The problem is that there is not much evidence for public outcries during the 1960s and 1970s in the newspapers or correspondence received by the Forestry Commission. The reality of sustained public opposition to forestry in the 1960s is probably a perception among foresters that helps them to justify their actions after the event. This lack of opposition to forestry in Scotland illustrates that we cannot always take oral testimony at face value and that cross-referencing with other sources is vital. In this case, the reality is that foresters themselves believed that the provision of recreational facilities and landscaping was a good thing, a view not driven much by outside pressure. In 1967 the Forestry Commissioners circulated an internal memo reminding all foresters that although public recreation was now regarded as important, it would always be secondary to the primary aim of growing timber.[28] The primacy of timber production was so ingrained into the culture of the Forestry Commission that it was hard for individual foresters to step away from it. That reflex was visible in the interviews when foresters were pressed on the question of whether economics and recreation were two conflicting objectives in forestry. Most foresters referred to the concept of multi-purpose forestry, the management of forests for more than one purpose, such as recreation and wildlife conservation, while preserving the long-term yield of wood.

As a result, the commission opened many of its forests to the public and even hired a landscape architect, Sylvia Crowe.[29] Her task was to make the public forests more attractive in appearance without interfering with wood production.[30] But that was not appreciated by all foresters, as Fred Donald acknowledged, "it is gardening on the site if compared with the bigger picture. It is the Silvia Crowe thing. She was a sensible woman, don't decry Silvia

Crowe because she acknowledged the need for forestry to be profitable." But landscaping and the ideas of Silvia Crowe was not applied to the new forest plantations in the remote north of Scotland and this left the Forestry Commission open to criticism with respect to the appearance of their plantations. The numbers of visitors to the forests continued to increase during the 1960s and 1970s. However, if there was any criticism from the general public it did not concern the Forestry Commission and their foresters much.

The real crisis in forestry emerged in the 1980s, when the concerted effort of conservation organizations, the Nature Conservancy Council (NCC),[31] and locals exposed forestry operations destruction of large areas of the Flow Country, a rolling expanse of ecologically valuable peat wetlands in northern Scotland.[32] Jim Atterson pointed out that this opposition "took them by surprise because we had been planting . . . from the mid-1960 onwards on deep peat." The opposition to planting in the Flow Country in northeast Scotland took foresters by surprise because for decades conservation organizations had largely agreed on the afforestation program. The leadership of the Forestry Commission were involved in organizations such as the Association for the Preservation of Rural Scotland (APRS) or the Royal Society for the Protection of Birds (RSPB).[33] Besides, the forester corps of the Forestry Commission was largely educated at the same universities and forestry schools and shared a common outlook and approach to forestry. For this reason, many of the discussions with regard to conservation, recreation, and other issues remained internal and largely invisible to the outside world. By the 1980s, a new generation of land managers and foresters graduated from newly established environmental and ecological departments and entered conservation organizations and government departments such as the NCC and challenged the status quo.

The old guard was not prepared for the influx of graduates with a background in environmental science and the fact that the new generation of foresters and landscape managers talked to local people and visitors of the forests. The new generation also carried out independent research on the impacts of forestry operations and found the work of the Forestry Commission environmentally wanting. In addition, the political lobby and a well-orchestrated campaign of conservation organizations and NCC against forestry operations in the Flow Country was so effective that the government had no choice but to shut down the planting program in northern Scotland. This took many state foresters by surprise because they believed that "[they] were doing a great job . . . creating a forest where there had been degraded peat

A monoculture forest plantation of Sitka spruce blankets the hillside of a Scottish glen. The crawler tractors on the right are plowing a fire trace. *Courtesy University of Aberdeen*

lands."[34] The new generation of environmentalists and land managers disagreed with the foresters and pointed out that the peat lands were irreplaceable and unique habitats and ecosystems. Some environmentalists regarded foresters as the villains in this saga, but history is rarely black and white. Most of those interviewed recognized that such historical developments are complex. Jim Atterson, for example, observed that value is in the eye of the beholder: "the actual creation of a diverse ecosystem had no value . . ., so the country should pay for it." It is true that from the 1970s onwards, public opinion is in favor of putting more emphasis on environmental issues and the protection of nature. The creation of healthy eco-systems and protecting biodiversity were increasingly regarded as valuable and worthy to spend public money on, facilitating the "ecological-turn" in forestry. This shift from the mass production of timber towards more emphasis on nature conservation, protection of landscape values, mitigation of climate change, and provision of recreation was largely facilitated by the long-standing interest in these issues by the foresters who created the Scottish forest plantations in the first place.

But we must approach these oral testimonies with some caution. There are hints that these foresters justify themselves after the event by appearing greener than they really were at the time. Still, without these oral testimonies, it would have been much more difficult to uncover the interest of foresters in landscape and environmental issues long before the modern age of environmentalism. Oral history reveals that many of these interests in wildlife, ecology, and landscape had been dormant amongst foresters but were obscured by official forest policy and the fact that only the poorest upland areas were available for forestry. The latter meant that environmental conditions limited what foresters could do with these lands. The interviews brought out the foresters' pride in what they have created and how good they became at cultivating and planting the land. Most of the Scottish upland moors were unsuitable for forestry and this forced state foresters to develop the techniques needed for cultivation and planting these areas from scratch. This successful planting laid the foundations for the new forest ecosystems that we see developing today. It is this very process of pitting oneself against a difficult environmental reality and the work done to shape these environments that gave foresters an intimate knowledge of the natural world.[35] The vision of these foresters was for the long term, to create forests for the use and enjoyment of future generations and we must applaud them for that. Oral history allows us to tell these stories of the real characters that shaped the forests and landscapes they managed. It illuminates the perceptions among foresters of the environments and the woodlands that they managed and how this is shaped by their education and institutional culture. It brings to life the interactions between these individuals and the environment and reveals what is hidden from the formal bureaucratic record of forestry organizations and government departments.

The official archival record often fails to document how foresters view their relationship to the forest or feel about environmental change caused by their actions. For example, there is abundant evidence in the popular press and campaigning material of what conservationists and civil servants in the NCC thought about the work of the Forestry Commission, which they regarded as destructive.[36] However, what working foresters thought about this assault on their work is obscured by the way that the Forestry Commission defended its approach to forestry since the early twentieth century in the official record. This lack of archival information prompted me to structure the interviews to tease out how foresters viewed the forests they helped to create and their attitudes to the natural world and conservation.

Among environmentalists in Scotland there has been a tendency to present foresters as environmental villains who cling to land management practices that viewed forests as production units of raw materials. By contrast, the oral testimonies I recorded reveal a more nuanced picture of adaptation and adjustment, shaped by the experiences of creating and managing forests and a deeply rooted interest in the natural world. Indeed, as a forest historian, one has to be wary of adopting wholesale declensionist narratives of modern forestry and forest exploitation, and to point out that contemporary perceptions of short sighted foresters employing destructive forest management regimes have turned out to be simplistic when analyzed against the oral history evidence.[37] As Jim Atterson explained, he and many other foresters went into forestry because of "the whole range of what is now called conservation," and that attitude partly explains why the ecological turn of forestry happened so fast in Scotland because it came from within the Forestry Commission.

The oral testimonies of these foresters not only speak for themselves but for the entire forestry community, the Forestry Commission and, above all, for the forests they created. These forests will remain long after their creators have gone and are a testimony to their dedication, knowledge, and skill. However, if not recorded, the stories of how the modern Scottish forests came into being and the motivations why these were created will be forgotten. Recording the recollections of the people who created these forests will make sure that we and future generations will understand how they created these forests and how these foresters felt about it.

Notes

1. A monoculture forest plantation is a planted forest that consists of one type of tree.
2. The conversion of bare land into forest is called afforestation.
3. Frank Fraser Darling (1903–1979) was an English ecologist, conservationist, and author. He is strongly associated with the Scottish Highlands and Islands through his publication on the ecology of Scotland. See, in particular, Frank Fraser Darling, *West Highland Survey: An Essay in Human Ecology* (Oxford: Oxford University Press, 1955).
4. The Flow Country of northeast Scotland is the largest expanse of blanket peat in the northern hemisphere. For the so-called Flow Country debate see: Charles Warren, "'Birds, Bogs and Forestry' Revisited: The Significance of the Flow Country Controversy," *Scottish Geographical Journal* 116, no. 4 (2000): 315–37.
5. Don MacCaskill (1919–2000), forester and naturalist.
6. Mark Anderson (1895–1961) was one of Scotland's most influential foresters during the twentieth century. He pioneered forest ploughing and advocated forest ecology. Anderson was professor of forestry at the University of Edinburgh.

7. Wilfred E. Hiley (1886–1961), professor of forestry economics, University of Oxford. See: Wilfred E. Hiley, *The Economics of Forestry* (Oxford: Clarendon Press, 1930).
8. Bob Allison was chief instructor for the Women's Timber Corps for Scotland during the Second World War. From 1946–1950 he was foreman in charge at Auchtermuchty. From 1950 until his retirement in 1970, he worked in the Speymouth forests and was appointed head forester in 1953.
9. Judith Tsouvalis, *A Critical Geography of Britain's State Forests* (Oxford: Oxford University Press, 2000).
10. Wood was used in mines for so called pot props. These were large wooden beams used to support the roof of a coal mine.
11. A Scottish forester who worked for the Forestry Commission as a researcher and district officer. He invented tree shelters sometimes referred to as "Tuley Tubes."
12. J. A. B. MacDonald (1900–1995), Forestry Commission siliviculturalist and Conservator of South Scotland.
13. Dame Sylvia Crowe (1901–1997) was an English landscape architect and garden designer.
14. For a detailed historical overview, see: K. Jan Oosthoek, "The nature and development of the forests since the last Ice Age," in K. J. Oosthoek, *Conquering the Highlands: A History of the Afforestation of the Scottish Uplands* (Canberra: ANU Press, 2013), 11–31.
15. *First Annual Report of the Forestry Commissioners* (London: His Majesty's Stationary Office, 1920), 11.
16. Ministry of Reconstruction, *Reconstruction Committee. Forestry Sub-committee: Final Report (Cmd. 8881)*, London: HMSO, 1918, 3–5.
17. For an in-depth discussion of the creation and subsequent history of the Forestry Commission see: Joshua West, "Forests and National Security: British and American Forestry Policy in the Wake of World War I," *Environmental History* 8, no. 2 (2003), 270–93; K. Jan Oosthoek, "The Origins and Development of State Forestry in the United Kingdom," in, *Managing Northern Europe's Forests. Histories from the Age of Improvement to the Age of Ecology*, ed. K. Jan Oosthoek and Richard Hölzl (New York: Berghahn, 2018), 130–65.
18. During the nineteenth century, Scottish landowners made conscious efforts to improve agriculture in Scotland. This resulted in the enlargement of fields, the introduction of new agricultural techniques, and the clearing of the uplands of people (so-called Clearances) and fields to make way for sheep.
19. For a discussion of an ecotechnological landscape, see: Robert Gardner, "Constructing a Technological Forest: Nature, Culture, and Tree-Planting in the Nebraska Sand Hills," *Environmental History* 14, no. 2 (2009): 275–97.
20. "Can't see the wood for the trees" means that someone is too focused on details to see the complete picture. In this case, it was assumed that foresters were so focused on creating single tree species plantations to produce timber that they could not see that these plantations are part of a woodland ecosystem.
21. Mather, A. S., "Forest transition theory and the reforesting of Scotland," *Scottish Geographical Journal* 120, no. 1 (2004): 83–98.
22. For a discussion of the "ecological turn in forestry" see: K. Jan Oosthoek and Richard Hölzl, "National Histories, Shared Legacies State Forestry in Northern Europe in Comparison," in Oosthoek and Hölzl, *Managing Northern Europe's Forests*, 359–89.

23. For a discussion of the problems related to upland afforestation see: K. Jan Oosthoek, "The Upland Question," in Oosthoek, *Conquering the Highlands*, 51–70.
24. Don and Bridget MacCaskill, *Wild Endeavour* (Glasgow: Blackie, 1975).
25. Rennie McOwan, "Don MacCaskill," *Herald Scotland*, May 16, 2000.
26. Declensionist environmental history are stories focusing on decline, where humans have degraded, damaged, or ruined the environment.
27. Mark McLaughlin, "Counterbalancing Declensionist Narratives in Environmental History," *Network in Canadian History & Environment*, February 3, 2016, https://niche-canada.org/2016/02/03/counterbalancing-declensionist-narratives-in-environmental-history/.
28. Robert Lambert, "'Therapy of the Green Leaf:' Public Responses to the Provision of Forest and Woodland Recreation in Twentieth Century Britain," *Journal of Sustainable Tourism* 16, no. 4 (2008): 415.
29. English landscape architect and garden designer Dame Sylvia Crowe (1901–1997).
30. David Foot, *Woods & People: Putting Forests on the Map* (Stroud: The History Press, 2010), 101–2.
31. Nature Conservancy Council (NCC) was a United Kingdom government agency responsible for designating and managing National Nature Reserves and other nature conservation areas in the United Kingdom between 1973 and 1991. In the mid-1980s the NCC carried out a peat-land survey of northern Scotland that provided the scientific evidence to prove the unique nature of the Flow Country ecosystem. This survey provided the ammunition that conservationists needed to convince the government to halt forestry operations in Northern Scotland.
32. Charles Warren, "Birds, Bogs and Forestry Revisited," 315–37. The Flow Country of Caithness and Sutherland is habitat for many unique and rare communities and species of specialized flora and their associated fauna, in particular, birds.
33. Oosthoek, *Conquering the Highlands*, 109–11.
34. Comment by Jim Atterson.
35. On the struggle between people and the environment and how it influences our knowledge and view of nature, see: Richard White, "Are You an Environmentalist or Do You Work for a Living?" in *Uncommon Ground: Toward Reinventing Nature*, ed. William Cronon, 1st ed. (New York: W. W. Norton, 1995), 171–85.
36. See: D. A. Stroud, T. M. Reed, M. W. Pienkowski, and R. A. Lindsay, *Birds, Bogs and Forestry*, ed. D.A. Ratcliffe and P.H. Oswald (London: NCC, 1988); David Goode, "Forestry and the Flow Country: Paradise Ploughed – A Television Programme in ITV," *New Scientist* 1598 (February 4, 1988), 85–7.
37. The danger of inaccuracy of many declensionist environmental history narratives has been discussed by Jon Sandlos, "Environmental History, Conservation, and the Social Sciences," *Network in Canadian History & Environment*, January 16, 2017, http://niche-canada.org/2017/01/16/environmental-history-conservation-and-the-social-sciences/.

Epilogue: The Fall and Rise of Oral Testimony in Environmental History

Christopher Sellers

When I was becoming an environmental historian in the late 1980s and into the 1990s, oral history lay far from the minds of the field's leaders. The path-breaking and field-forming works of this period sought historical perspectives on entire ecosystems and on distant origins that were usually too remote for contemporaries to remember. More macro perspectives and plunges into the long past reflected a shared resolve not just to sharpen the field's tools but to professionalize it, to make it a field in which historians could make their careers, in which history department hires could happen.[1] With a few exceptions, such as Samuel Hays, environmental history's founding mothers and fathers saw their work as broadly illuminating the origins of environmental crisis that were distinct from environmental politics or advocacy, and thereby qualified as bona fide *history*. And within American history, where most staked their early claims, they were determined to distinguish their historical enterprise from other subfields, political and social as well as intellectual history. In this endeavor, a personal and individual witnessing of past human–nature relations seem too precarious an evidentiary foundation for that "nature" whose human history they sought to write.

Instead of testimony, these early environmental historians trusted far more in contemporary science as their new field's epistemological rock. Ecology in particular was taken as the gold standard for elucidating nature's past contours and dynamics. Researching and writing my dissertation as I read all this pioneering work, I sought to add sciences of occupational and environmental health to the mix, but shunned interviews with elderly architects of these fields.[2] Mistrusting the biases and other distortions of memory, I was convinced that all I really needed lay in published studies or in the archives. Not surprisingly, I didn't crack the cover of the *Oral History Review* until the late 1990s.

Christopher Sellers, *Epilogue: The Fall and Rise of Oral Testimony in Environmental History* In: *Oral History and the Environment.* Edited by Stephen M. Sloan and Mark Cave, Oxford University Press.
DOI: 10.1093/oso/9780190684969.003.0014

I look back now through the earliest issues of the journal *Environmental History* and its more scholarly predecessors over these same years, and see I was not alone.[3] My own decisions conformed to the larger patterns consolidating within the new academic field of environmental history. To tell by their mentions of "oral history" or "oral histories," articles in *EH* and its scholarly precursors downplayed oral history until the late 1990s, with most mentions confined to titles in endnotes.[4] Oral historians, at least judging from that field's professional journal, returned the favor. Over nearly three decades from 1973 to 2000, the *Oral History Review* only ran two articles and a single review even that mentioned "environmental history," with "social history" drawing vastly more references.[5] But this mutual disregard between oral and environmental history was much less evident in the *Journal of Forest and Conservation History* (*JFCH*), a precursor to *Environmental History* with less academic origins. From the 1950s, *JFCH* had published oral histories of leading figures in the profession of forestry and from paper companies like Weyerhaeuser, to paint a mostly self-regarding portrait of this profession and industry's past. In the mold of the elite oral history favored by those such as Allan Nevins,[6] its interviews yielded a history largely composed by and for foresters themselves, along with the private timber industry for whom they so often worked. Even after editors also began to interview Sierra Club members and fold in more pieces by trained historians, this journal's continuing reliance on oral histories reflected its less academic status and audience, even as its competitor—the *Environmental Review*—gained in scholarly credibility and readership.

By the late nineties, when oral history began piquing the interest of more environmental historians, the door between the two fields was cracking open through the new relationships being forged between environmental and social history. Aiding the integration of the new field into history departments, a new fusionist socio-environmental history brought, among other insights, a new critical take on the conservationist project, for its consequences on rural and indigenous peoples.[7] Environmental historians also trod well beyond the forests and farms to take on industries, cities, and health issues that had been such central concerns of the post-World War II environmental movement itself. Leading advocates of environmental history's growing engagement with social history nevertheless drew reassuringly sharp lines between their own integrative project, now rooted in social as well as ecological science, and cultural imaginings of their historical actors. When Ted Steinberg in 2002 urged a "concern with power and class" in environmental history, he

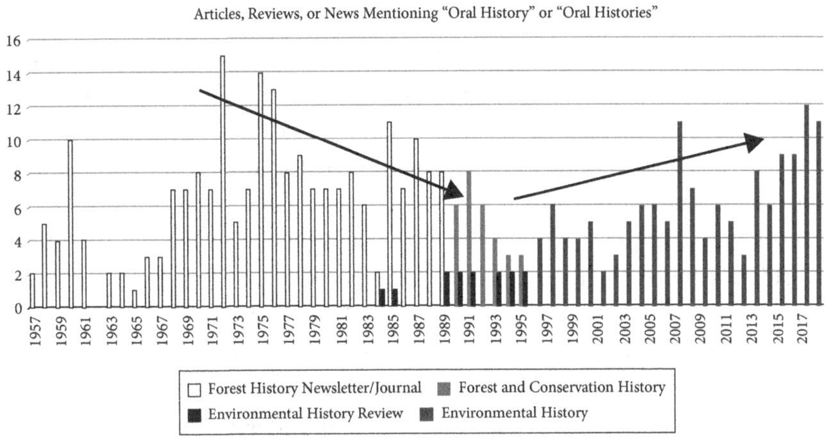

This graph tracks mentions of "oral history" in American environmental history journals. *Graph by author*

framed that direction as directly opposed to concerns with "the social construction of knowledge," which amounted to "looking over their shoulders at the postmodern ghosts chasing them."[8] Even an environmental historian as welcoming to cultural history as Richard White found the terrain explored through oral history to be one of "memory," distinct and separate from that of professionally composed "history," environmental and otherwise.[9]

References to oral history nevertheless picked up through the first decade of the new century and especially during the 2010s. Even more significantly, more of those mentions appeared not just in footnotes but in articles, as oral history has come under discussion for how it may contribute questions and substance and bring added complexity to environmental history as an intellectual enterprise.[10] The momentum of this shift suggests that it has been favored not just by scholarly bridge-building but by more tidal and contextual change.

One influence is the ever-further extension of environmental history beyond North America and Europe, to better encompass the Global South. As the field's center of gravity shifts away from where not just environmental history but the history profession itself first arose, it has confronted the challenge of uncovering pasts that for these parts of the world are less documented or archived in the classic Western sense, except by colonizing powers. Faced with such limited and patently slanted written records, scholars have been turning to oral traditions to provide additional windows into past environmental change. More subtly, this collection suggests that the further impacts of environmental history's global turn may prove even more far-reaching. As oral environmental history has been not just been revived but gained a more central place in historical inquiry, what Steinberg dismissed as "postmodern ghosts" are turning ever more substantial, disrupting some of environmental historians' most long-standing and cherished assumptions.

Colonial and post-colonial conflicts erupting across these parts of the world have increasingly heightened the imperative for our field to, as Dipesh Chakrabarty put it a couple of decades ago, "provincialize Europe."[11] By that phrase, he meant not just extending our scholarly gaze and tools to other parts of the world, but using vantage points we find there to reexamine our own Europe-derived ways of seeing, knowing, and understanding. Chakrabarty offers the example of an 1855 rebellion against British colonizing forces spearheaded by a tribe acting on orders of their god Thakur, whose reality and role were then shrugged off both by the British and historians of this event. Since Chakrabarty's book was published in 2000, scholarly appreciation has been deepening of how Western state power and those sciences and worldviews nourished within it are not automatically universalizable. For indigenous as well as many other less powerful groups, their adoption or appropriation has proven far from inevitable. As we are learning, that is often because for such groups, they offer insidiously disempowering vantage points on the natural world and humans' place in it.

Thus far, those academic fields carrying this kind of critical analysis forward have included an environmental humanities ever more engaged with how culture and identity shape people's ideas and experiences of the natural world, also environmentally minded oral historians. Together, they are opening exciting new scholarly borderlands that others in environmental history should welcome. By taking the words and worldviews of non-scientist social groups more seriously, environmental history may finally be coming to grips with an insight that is far from ghostly: how pervasively power differentials structure not just social conflicts but contrasting ways of seeing, understanding, and navigating the natural world. The work of oral historians in this collection makes plain how different cultures and social positions actually do yield fundamentally different framings and experiences of the environment and one's place in it.

Arguably, this new interest in oral history among environmental historians also reflects the growing corrosion of expertise and scientific authority along many fronts, as digital interconnectedness and a proliferation of new media have fragmented the ways we speak about nature and the environment. While this corrosion has famously been exploited by an anti-environmental right, these essays suggest just how limiting and distortive it is to frame the politics of science solely through a dichotomy of an "anti-science" right and a pro-science left. Instead, they explore the sheer variety of ways power imbalances and conflict weave through questions about what nature and environment are, who gets to speak about them, and whose voices come to be heard. For environmental history, a field that has long striven for independence from environmental advocacy, we learn here just how shot through with politics even our own customary pathways through these realms can be, leaning as they do on today's sciences.

This collection of essays outlines contributions of oral history and oral historians that go well beyond those enumerated by Harold Steen and Susan Shrepfer, in a 1972 issue of the *Journal of Forest History*. Nearly fifty years ago, they emphasized mainly the "valuable supplements" that oral histories added "to conventional sources," by revealing "personal beliefs and rumors as well as the actual events in which" interviewees took part, also by providing "intimate views of incidents that make for dramatic, colorful, written histories."[12]

Here in the second decade of a twenty-first century, this collection's contributors wield oral history in ways that are more intellectually programmatic and enriching. It opens windows onto connections that documentary sources miss, and doors into histories that environmental historians' customary assumptions and methods slight. The oral testimony explored and analyzed across these chapters suggests just how deep and thorough are the differences in what different historical actors take nature or their environments to be, what a philosopher would term their "ontology," the things and settings they understand to make up their world. Throughout this collection, contemporary science—long environmental historians' go-to window on what nature actually is—repeatedly reveals weaknesses and blind-spots. These are highlighted precisely by authors' efforts to dig out other ways of knowing, contrasting ways of seeing and interpreting nature's as well as humanity's realities, that "technocratic prejudice" has often led scientists or officials to dismiss. These essays suggest that for environmental historians more fully to engage with how power manifests itself in history, instead of simply averting our eyes from "postmodern ghosts," we'd do better by staring them down For that, oral history can be valuable tool, providing essential insights. These essays collectively argue there remains a need, even an imperative, for listening *harder* to the voices and accounts of ordinary people.

Among those "mono-truths of history and reality" that oral history can aid environmental historians in better contextualizing and diversifying are characteristically modern and Western assumptions about time. As historians are now finding, the type of time that governs science as well as most of professional history today is one of calendars and clocks that first emerged in Western nations beginning in the eighteenth century.[13] By and large, environmental history assumes a passage of time universal in its pace, and in particular, place, such as a community or nation-state defined by periods or developments that are shared by all historical actors. But these senses of time, or temporalities, can match up poorly with the more diverse temporal experiences revealed through oral histories. This collection explores those many alternative temporalities supporting how people understand, respond to, and cope with environmental continuity and change.

Among the most important ways these oral histories open up experiential alternatives to the clocks of contemporary science is through testimony that is not just personal but also more or less explicitly communal, thereby reaching beyond the time horizon of any individual, invoking cross-generational connections and continuities. This is especially striking in the three of the contributors' research that centers on indigenous groups: Jan Bender Shetler (people living in the Mara region of Tanzania), Caren Fox (the Māori of New Zealand), and Tero Mustonen (the Evenk in the Eurasian taiga). In each, cited storytellers describe group knowledge and practices honed stretching back generations that connect them not only to their own forebears but also to the environments these groups have inhabited for so long. For instance, the geothermal springs of New Zealand were well known and much used by Māori, figuring heavily in their cosmology and spiritual lives. By contrast, the recent, accelerating industrial takeover of the geothermal springs, along with the destruction of the taiga, have threatened more than just targeted "natural resources." Both today's industrializers and modernizers also threaten indigenous ways of knowing and living through time. Like earlier colonizers such as the British (in the case of Tanzania and New Zealand) and the Russians (in the case of the taiga), these latter-day colonizing forces have demonstrated little interest in recognizing any temporality other than their own. Oral environmental history, by contrast, can illuminate and help preserve indigenous versions of time.

A communally shared, multigenerational time sense also can develop among peoples whose habitation of land has not been long-standing enough to qualify as "indigenous." Among the architects and maintainers of community-based and highly local systems of water supply and distribution in Puerto Rico, Javier Arce Navarro's oral histories reveal a decades-long trajectory and sense of accomplishment starkly contrasting with assumptions made by state officials and experts. For public health officials, connection to a large-scale water system constitutes environmental improvement, a clear metric of developmental progress. Yet, as Navarro shows, contrary tales exist among those who nevertheless pursue more local communal management of their water supply. Their accounts suggest that in at least one sense, alternative "backward" methods actually work better in health officials' own terms of assessment: it is more adaptive to drought, perhaps also other kinds of environmental duress. Other of its virtues for locals themselves, however, escape public health officials' purview. For instance, local control fosters a self-determination and collective agency they would not enjoy in a centralized, technocratically administered water system. Testimony about drought gathered by Deb Anderson from settler narrators living in Australia's wheat belt also extends beyond individual efforts to encompass their predecessors' struggles and ability to cope. The multigenerational memory of similar struggles to farm an episodically dry land reaches back as far as a century or more. Stories of endurance across these generations serve as a shield for coping with current uncertainties: a source of solace and hope often conjoined with grim humor.

While one set of alternative temporalities explored here comes from speaking with those living a long time in a single place, other oral histories here follow the temporal trajectories of those who travel or migrate. Long inclined to narratives centering on particular places, environmental historians have ventured little into the history of human migration until recently.[14] Oral history usefully illuminates the human movement stemming from environmental change. Nowhere are these migratory consequences more striking or important than in environmental disasters. As Heidi Hutner's interviews of Fukushima survivors show, this disaster's scope was not confined to either the land surrounding that nuclear plant, nor even to wide dissemination of radioactive particles into food chains, ocean, and atmosphere. Fukushima also sent residents fleeing, splintering families and fueling a host of conflicts with an officialdom that has sought to downplay or deny the scope of the disaster.

Conversely, Mark Cave's interviews of those involved in relief efforts after the BP oil spill track how media coverage drew in volunteers from far outside the zone of ecological damage. Traveling in from distant points, their senses of trajectory intermingled with those of people mobilizing locally to assemble a more provisional temporality of collective response. A similarly converse dynamic of a disaster's historic impact unfolds in Timothy Cooper's and Anna Green's oral history work on the *Torrey Canyon* oil spill. There, the impacts of the disaster were magnified far beyond the site itself through the media coverage and in retrospective histories positing it as a spark for the environmental movement in Britain. But Evans, their interviewee, remembers it as more of blip, a "fragmentary" event that only seemed a relatively minor step in her own pathway toward environmental activism. Oral histories of on-the-ground witnesses such as Evans pose questions about the purportedly seminal role of the *Torrey Canyon* incident in the birth of an environmental consciousness and politics in Britain.

Suroopa Mukherjee, by examining the use of oral histories among Bhopal activists, suggests that oral history-taking itself can extend the temporality of a disaster beyond its ground zero, as well as its materio-ecological extensions, in ways that serve to empower its victims. As Mukherjee recounts, as is also evident in the public testimonials of Hutner's witnesses, creating those spaces and social settings where those who had been impacted could tell their stories itself extends the story of the disaster in critical ways. The very act of recollecting the resulting traumas perpetuates their visibility, keeping them alive. Witnessing asserts "entitlement" and otherwise pushes back on those many forces and interests seeking to obscure, confine, or erase the impacts of Bhopal or Fukushima by consigning them to a past that is becoming ever less relevant. Talking about the disaster and the ruptures it has brought to the lives, health, and livelihood of so many serves to remind, to keep the trauma suffered by a disaster's least fortunate of victims visible and audible. Oral histories of these disasters shed a more honest and inclusive light on the long-term and continuing face of disasters that strike suddenly but then recede from the headlines and most people's awareness.

The research presented here also calls attention to the ways in which oral history can open up avenues beyond the debate over narrative declensionism in environmental history—the field's fixation on stories of environmental destruction and "decline"—launched by William Cronon nearly three decades ago.[15] Defenses of declensionism aside, environmental historians have since forged many non-declensionist approaches to environmental transformation and change. Often, these narratives hew more closely to the evolving experiences and practices of people who continually inhabit the places under historical study.[16] As recently noted by the oral environmental historians Katie Holmes and Heather Goodall, oral histories can propel such projects forward by asking (among other things) "what does it mean for environmental history to take seriously the embodied and affective nature of memory . . . ?"[17] Years of living in a place, imbibing its sights, sounds, and smells through one's own body, can impart not just knowledge but lasting affection for that place, even among historical actors who have had a hand in its destruction. Hence, Osthoek's foresters—classic official agents of conservationism in much of the literature—tell stories that run counter to what he explains as a classically declensionist trajectory of the efforts of Scottish forestry on the Highlands over this period, as compelled entirely by extractivist and profit-minded ambitions. Foresters' own sense of this same passage, his summary and analysis suggest, reveal many contrary impulses, friendlier to aesthetic and recreational values.

In better exploring and appreciating how multiple and contradictory the temporal trajectories of historical actors can be, historians may also better capture just how deeply the experience of environmental transformation or disaster can diverge between the privileged and powerful and their less fortunate contemporaries. Not only do the temporalities of indigenous or long-standing residents often contrast with those of state or corporate officials, social divides by gender and class can undergird substantial differences in the pace and direction of disaster's course. Among several of these case studies, women's sense of time's passage flows differently than men's. Around Lake Victoria, women's stories about disastrous droughts followed expanding social networks between tribes, with a pacing that was distinct from the thread of consolidating ethnic identification emphasized by men. For women whom Fukushima turned into unexpected feminists and activists, goals taken up included resisting efforts by the government to whitewash continuing risks and otherwise neatly tuck this disaster into Japan's and the world's past. As they set out to testify and otherwise publicize what they continued to endure, however, they ran up against Japanese culture's prevalent silencing of women. While some left to join expat communities, others like Sachiko Sato stayed in Fukushima to resist, her newfound feminism indelibly interwoven with her resistance to official efforts to downplay and deny the continuing hazards. Cooper's and Green's interviews on the *Torrey Canyon* disaster also suggest that a gender divide figured into people's personal genesis into environmental activists. Whereas a masculinist narrative was more apt to link this spill to modern environmental awareness in direct linear fashion, interviews with British women activists show the disaster to have been considerably less influential or directly transformative.

Similarly, a class-based contrast in temporal experiences of the Bhopal disaster has complicated the dilemmas faced by its most aggrieved survivors. Lower-class residents of Bhopal lived nearest to the plant and suffered its harshest impacts. Subsequently, while the disaster is by no means over for them, they've struggled to ensure that their own experiences in that horrendous moment and beyond are not forgotten. As the years have worn on from 1986, much of India's officialdom as well as its middle class seek to move on, increasingly relegating the disaster and its impacts to their pasts. In response, those for whom the disaster is by no means over, in whose own bodies and lives it continues to play out, have had to mobilize, to push an alternative temporality in which the disaster's traumas continue to be voiced and heard. Taking on the tidy and purportedly universal narratives for such disasters preferred by an officialdom with their discretely drawn endpoints, victims' groups have been undertaking political struggle on the battlefield of time itself, through public insistence that this disaster has by no means clocked out.

Perhaps the furthest-reaching challenges to environmental historians' usual assumptions about time come in the work of Cooper and Green, and of Lee. For the former, "continuous" time stands in contrast to a "local, discontinuous representation" of time's passage, what they call "fragmentary time." For Lee, the more or less "linear" trajectory embedded in most environmental (as well as other) historians' textual narratives stands in contrast to the "freer" and "wilder" leaps she finds in oral testimony. These important distinctions help clarify key tensions between oral history-taking and historians' traditional ways of composing their own narratives. Given environmental historians' characteristic emphases on chronologies and time flows defined by the sciences, these tensions are sharper for them than those working in many other kinds of history, hence more deserving of attention in the planning and execution of oral histories on environmental topics. As Lee suggests, environmental historians undertaking interviews should take care to hold in check their own pre-established notions about the flow of relevant events. Instead, by avoiding leading questions and patiently listening, they need to leave room for the historical subjects being interviewed to explore *their* sense of the timing, linkages, and significance of remembered events deemed relevant. As Cooper and Green demonstrate, once we open our ears to the time senses of our oral interviewees themselves, we may discover new perspectives on our inherited narratives about historical change. In my view, acknowledging a certain freedom and fragmentation in the temporalities of our oral narrators does not require an abandonment of other temporalities defined by science or by the guardians of traditional historical method. Instead, taking the fusionist project of a socio-environmental history a step further, the work of oral historians in this volume calls for us to interweave our traditional approaches to time with the ways less privileged groups experience time's passage. These essays also call for environmental historians to pay more attention to those worlds to which less empowered historical actors give voice—their alternative visions of landscapes, nature, and other of the very "stuff" within which they dwell.

Beyond their challenges to the conventional ways of incorporating time into environmental history, oral history also suggests alternative ways that natural worlds or environments themselves can figure into historical narratives. Ever since the 1970s and 80s, ecology and other contemporary sciences have served as mainstay for providing the non-human nature that environmental historians incorporate into their histories. That commitment, equating "nature" with what today's scientists say it is, shored up resistance in some corners of this field to insights about the constructed-ness, contingency, and power relations supporting natural science, as revealed by social and historical studies of scientists and their work. While the field has since become more open to and integrated with these neighboring fields, a relatively settled consensus about environmental history as a "materialist" enterprise still leans heavily on the lenses of contemporary science.[18]

The growing turn to oral history by environmental historians amounts to a counter-current, opening the field to less science-centered ways of grasping the "stuff" or being(s) of nature. It parallels similar trends across many other fields, from an "ontological turn" in science studies and philosophy to explorations of "alter worlds" in ethnohistory to a program of "political ontology" in political science.[19] Today, when oral historians Debbie Lee and Katherine Newfont argue in their edited collection that "the land" itself "speaks," they echo Leopold's "thinking like a mountain" essay but also suggest something more.[20] Newfont's and Lee's "fluency" of the land does not require a scientific professional to understand; it is less hinged than was Leopold's essay on the science of ecology itself. For them, and that volume's contributors, as for many essayists here, the land's language is discernible, instead, through the perceptions, categories, and testimony of those living on and working this land.

Essays in this collection suggest that an ontological turn may already be underway at the intersection of oral and environmental history, coming precisely as oral historians take on environmental projects and as the field itself is taking oral history methods more seriously. Even those most closely hewing to the role science classically plays in environmental history—Jan Oosthoeck's history of Scottish foresters and Dai Qing and Kang Xue's account of the struggle against the Upper Yangtze Hydroelectric Project—wield oral history evidence that provides important counterpoints. Oosthoek's analysis unpacks an alternative ontology lost in a widely accepted narrative about the arrival of conifer plantations in the Scottish Highlands, to not so much overturn as complicate narrative. However profit-driven and destructive the overall effects of their work, foresters themselves recall commitments to sustaining these forests over the longer term, also to an aesthetic appreciation of this landscape on which they laid such transformative hands. Some were even able to carve out preserved, appealing corners of the woods that served tourists rather than timber corporations. In Xue and Qing's chapter, while expertise remains central to an NGO's effort to block the dam, by bringing out its ecological impacts on rare and endemic fish species, Zhang Boju and his Friends of Nature group rely even more so on getting word out publicly. Knowledge about the government plans and their ecological consequences thereby came to be shared far beyond the closed circles to which government officials would happily confine them. Nor is Boju's political opposition just about threatened fish species or forests; it is very much people-oriented. Friends of Nature sees itself as speaking and acting on behalf of a "silent majority" of the Chinese people, who "lack the education and social skills to recount their experiences" but who are forced to endure "the largest sacrifices made in the name of China's rise."

Several essays also reflect on the limits and drawbacks of that lens afforded by Western science and scientists on the environment ("nature") itself. Multigenerational narratives unpacked by oral historians can clash directly with contemporary science; their persuasiveness can override scientific or other sorts of facticity. As Anderson notes, trust in them may also make rural Australians resistant to messages that climate change may be making these droughts worse, hence less survivable. Or, as in the recurring stories told Jan Bender Shetler about children floating into villages on Lake Victoria during times of scarcity, they can soften remembrances of a harsh practice like the selling of children for food. Caren Fox's Māori interviewees have lived much longer around the geothermal pools that lie at this center of this essay than have the institutions and experts wielding science to understand and use them, which imposed damaging subsidence on a Māori ceremonial site. Even as the Māori now accept modern geologists' accounts and explanations of the geothermal fields, one narrator expresses reservations about one point where the science clashes with traditional Māori understanding. Though geologists insist these pools are not connected, Aroha Campbell still suspects they are, since "if you go down far enough you hit magma and it is all linked."

Oral histories open doors to dwelt-in worlds for which a science-based environmental materialism has little room, other than perhaps to consign to superstition. Most striking are the ways Māori sustain their traditional cosmology: by taking the reservoirs as "living . . . almost in the literal sense," by how their beliefs in "various gods and the ancestors . . . assume a central role in [their] management narratives," and by how some of their sacred mountains and other land features are coming to be recognized in national laws. These connections opened to more metaphysical realms are not just visible in the essays featuring indigenous groups. Some of those converging on Louisiana to help in the wildlife rescue in the wake of the BP oil spill, for instance, when asked, also voiced their sense that these animals had "souls," a metaphysical notion that remains common lingo across Western religions. Both for Westerner and indigenous peoples, the alternative ontologies unpacked by oral histories can help illuminate terrain often tucked to the side of environmental histories by being categorized as "religious," which nevertheless continues to shape historical experiences of the natural and human worlds. However secular our own society has become—a matter of debate—when environmental as other historians simply foist aside the faith-based worlds and worldviews of their historical actors, they replicate the moves of colonizing forebears. Seen in this way, projecting our own resolutely secularized ontologies onto worlds to which they are more or less alien is not exactly a neutral or objective act. Research presented here suggest such moves need to be handled, at the very least, with more care and sensitivity than our field has often shown.

Tero Mustonen's essay goes so far as to stake out a way of knowing nature among his interviewees—herders, fishermen, and farmers in the Finnish and Russian taiga—that is fully independent of natural science. Faced with stepped-up resource extraction in the wake of the Soviet Union's fall, they maintain varieties of "traditional ecological knowledge," translatable into modern ecological terms with relative ease. But Mustonen also argues for a further reaching alternative he dubs "indigenous knowledge," not so easily subsumable into "frameworks of North American [resource] management science." The Evenk, though in many ways a European people, had crafted what Mustonen terms "Earthviews" with respect to the taiga, a "diverse and rich understanding of the situation on the ground" further defined and honed as it was passed along, Oral history can get closer to this style of knowledge and even help perpetuate it, since its "materials are valid through their being spoken—and then documented—*events*, through which the traditional people, a community, or the individual have named and recounted their world and its issues." The capacity of oral history to fully bolster community-based Earthviews owes much, as Mustuonen's own methods suggest, not just to listening but to a careful incorporation of locals' agency and influence on the design of oral history projects.

Narrators highlighted through this research challenge not just the ways of knowing and seeing derived from natural sciences but the public and private institutional landscapes which they see themselves as inhabiting. Especially notable across the essays are the "technocratic" forces driving big infrastructure projects, which leave many lives and livelihoods in the shadows only to wind up trampling upon them. From the Upper Yangtze Dam project to the Puerto Rican water supply system to New Zealand's exploitation of geothermal power, officially steered or sanctioned "mega-projects" are cast into a new light through oral histories of those living along their edges. In the case of Javier Nazario's interviews with a manager of Doña Mayo's localized water supply system in Puerto Rico, we learn much about how, contrary to widespread assumptions propagated by state officials and public health professionals, it actually works. It does indeed deliver water with little risk of spreading disease, at least by this testimony, and is better able to sustain that supply in periods of drought. Among its other virtues, ignored by those still advocating for universal large-scale water supply, the local community actually runs this system itself, making it less beholden to neglect or faulty decision-making by officials in a distant city. Oral histories also illuminate how in New Zealand, the Māori successfully pushed Contact Energy to begin repairing damage it caused to their ceremonial sites and were even able to gain a share of the expanding industry of hydroelectric electricity production. They have done so, even coming to respect and trust Contact Energy's reparations, both by taking up these modern technologies and by incorporating a "sustainable" management ethic they see as consistent with traditional tribal values.

As with contrasting senses of time, the openness of this volume's authors to contrasting ways of knowing and engaging the natural world as well as institutional landscapes helps them highlight dilemmas faced by the less privileged, whether because of race, gender, or class. From the Global South to industrialized Britain and Japan, women in many cases occupy a different world than men: from recollections centered more on inter-ethnic relations (Tanzania) to more pointed concerns about lingering toxic dangers (Fukushima) to memories of an oil spill as a less-than-transformative moment (*Torrey Canyon*). And the traumas of Bhopal that continue to ramify among its lower-class victims no longer afflict a middle class and a regional and national officialdom, helping explain the latter's readiness to declare this disaster "over." On the other hand, those at the receiving end of powerful, damaging interventions can and do find ways of resisting such silencing. From the mobilizations of Bhopal victims to the Earthviews of Mustonen's taiga inhabitants, they too craft sociopolitical means for raising their voices, to assert their own ecologies, interests, and entitlements.

Through the lens of oral history, those experts who at first glance seem the prime agents of "monotruth," the scientific managers of "natural resources," possess greater complexity and contradiction than declensionist environmental history narratives usually allow them. While they may well be scientific or extraction-minded in their work, speaking with them at length about their memories and experiences can turn up alternative viewpoints and affections. Cave does find that appreciation of environmental suffering, in this case of birds in the wake of the BP oil spill, does differ between the volunteers and the scientifically trained professionals charged with the governmental response. But like Osterhoek's foresters, scientific professionalism here does not preclude backgrounds and personal storylines evincing emotive allegiances or attachments to those creatures they are charged with protecting. Thereby, even in the same person, significantly different and clashing perspectives on nature coexist with one another. As Debbie Lee's interview and analysis suggest, however scientifically trained, the guardians of our more classically natural spaces, wilderness areas like the Selway-Bitterroot, effectively incorporate these places in human-scale stories that deeply engage their memories and emotions. That contrasts sharply with how our culture and laws often define wilderness, by an absence of human interventions or imprints. Among the most haunting memories reported by one such caretaker is about the recovery of a co-worker's corpse in one of its rivers, a harrowing image that humanizes this place for him by reminding of how hostile to people it can be.

If oral histories of scientific professionals here evince a kind of ontological pluralism—bearing multiple, sometimes contradictory versions of the natural world—that multiplicity is even more apparent among those whose relationships with their environments are less mediated by the sciences or a state-based technocracy. Often lost in the large-scale or macro vantage points adopted by many environmental historians are, as Javier Nazaro puts it, the fact that often, "conservation and adaptation to environmental change is done by everyday people, at household and local scales." By giving attention and voice to this more local and intimate scale of human actions, we render more visible the experiences of those who remain closely connected to the water or other part of the natural world, who over decades or generations have honed their own ways of understanding and managing it. Through oral histories, environmental historians may find narrative alternatives to a "polarized battle between humans and nature," historical ontologies that don't pit the two against one another as inevitably at odds. Caren Fox's Māori weave their traditional understandings and cosmology into how they monitor and manage their own geothermal enterprise, for instance. Forging an ontological pluralism of its own, the tribe comes to run its own geothermal operations, infusing these with frameworks and obligations to these places drawn from its own sacred realities. The Māori are thereby able to come up with less destructive and more genuinely sustainable ways of using the springs as resources. Fox suggests it provides a model for other parts of the world: "listening to the voices of the guardians of their environment . . . those indigenous peoples . . . that live on their land . . . their values, combined with the expertise of developers may hold the key" to resource development that is genuinely sustainable.

What these and other oral historians of the environment have set about demonstrating is first, that the ecology whose histories we seek has indeed differed according to the eyes, mind, and memory of the beholder. Second, actually recognizing and unpacking those differences will not plunge the field of environmental history into a morass of "postmodern" relativism, but instead strengthen its collective grasp of how power operates, both within and through the histories we write. To ignore such differences, to continue simply to assume that today's science hands us the only viable and working definition of ecology or the environment, is to continue to ignore just how insidiously and pervasively power threads through nature's many roles in human history. Environmental historians would be better off confronting this variety of vantage points and conceptions among our historical actors head-on, just as our colleagues in fields from anthropology to geography to political ecology have already done. Already, the many alternative temporalities and ontologies unpacked here suggest that ongoing intersections between oral and environmental history are laying groundwork for a "pluriversal" approach to the human–nature relationship and its history. As the sociologist Janet Conway and political scientist Jareet Singh explain, "notions of the pluriverse imply multiple ontologies, multiple worlds to be known—not simply multiple perspectives on one world."[21] Whether or not our essayists here are willing to go that far, certainly the sum of their work points toward a more inclusive and even-handed sense of just what the ecological dynamics are in a given time and place, also of the potentially contrasting ontologies and temporalities through which these dynamics come to be apprehended and experienced. At the very least, this collection as a whole makes the case that distinctly different ecologies and time senses may coexist in the same moment and place, even in the same person, also how insistently power differentials determine the interaction among them.

On this question of power, many of these essays converge upon similar surmises. In so many instances, oral history seems to be coming into its own as a method in environmental history by plumbing all that is missed through a simple equation of nature with what contemporary science shows it to be. The revelations come through an opening of ears, minds, and analyses to those temporalities and ontologies that non-scientists—the vast majority of historical actors—themselves see and care about. Thereby, we also learn how ecologies and timing determined exclusively by natural scientists and state officials tend to slight especially those ontologies and temporalities experienced by less powerful groups. A common, larger narrative also seems suggested, as promising grounds for many book-length environmental histories to come. So many of the temporalities as well as ontologies described here have themselves only turned "alternative" over time, in confrontation with powerful forces that have sidelined those interviewees who wield them.

Contributions here exemplify a certain intellectual ferment that environment-minded oral historians are bringing to the field of environmental history. In the process, the usefulness of oral history for environmental historians has flipped around from that envisioned decades ago in the pages of forest history journals. Back then, it served mainly as supplement, fleshing out and humanizing narratives about forestry or a conservation movement largely taken as known. Now, oral history is extending and amplifying the very questions being asked by environmental historians and even suggesting new narratives, less centered around science-derived pacing and worldviews than those of ordinary people and groups. By taking oral history itself more seriously, environmental history can take their inquiries into the workings of power and social conflict to another level, one that better respects and probes how deeply experiences of human–environmental interaction may diverge and clash, both in substance and in timing.

Notes

1. See Donald Worster's essay and the responses by William Cronon, Alfred Crosby, Carolyn Merchant, Steve Pyne, and Richard White, in a 1990 *Journal of American History* Roundtable, beginning with Donald Worster, "Transformations of the Earth: Toward an Agroecological Perspective in History," *Journal of American History* 76, no. 4 (March 1990): 1087–1106.
2. Christopher Clare Sellers, "Manufacturing Disease: Experts and the Ailing American Worker," PhD diss., (Yale University, 1992). http://search.proquest.com/pqdtglobal/docview/304015779/abstract/84460C741F5346F3PQ/1.
3. Currently published by Oxford University Press on behalf of the Forest History Society and the American Society of Environmental Historians, *Environmental History* came from a merger of the journals *Forest & Conservation History* (1990–1995) and *Environmental History Review* (1990–1995). Antecedents of *Forest & Conservation History* are *Forest History Newsletter* (1957–58), *Forest History* (1959–1974), and *Journal of Forest History* (1975–1989). The predecessor to the *Environmental History Review* was the journal *Environmental Review* (1976–1989).
4. Thomas Cox, "A Tale of Two Journals: Fifty Years of 'Environmental History': And Its Predecessors," *Environmental History* 13, no. 1 (2008): 33.
5. Based on keyword searches for "environmental history" and for "social history" in the *JSTOR Arts and Sciences VII* database; "social history" drew 131 mentions.
6. Alistair Thomson, "Four Paradigm Transformations in Oral History," *Oral History Review* 34, no. 1 (2007): 51; Jerrold Hirsch, "Before Columbia: The FWP and American Oral History Research," *Oral History Review* 34, no. 2 (2007): 1–16.
7. Alan Taylor, "Unnatural Inequalities: Social and Environmental Histories," *Environmental History* 1, no. 4 (October 1996): 6. https://doi.org/10.2307/3985275; Ted Steinberg, "Down to Earth: Nature, Agency, and Power in History," *American Historical Review* 107, no. 3 (June 2002): 798–820. https://doi.org/10.1086/532497; Nancy Langston and William Cronon, *Forest Dreams, Forest Nightmares: The Paradox of Old Growth in the Inland West* (Seattle: University of Washington Press, 1996); Karl Jacoby, *Crimes Against Nature: Squatters, Poachers, Thieves, and the Hidden History of American Conservation* (Berkeley: University of California Press, 2001); Mark David Spence, *Dispossessing the*

Wilderness: Indian Removal and the Making of the National Parks (New York: Oxford University Press, 1999).

8. Steinberg, "Down to Earth," 803 ("ghosts"), 804 ("power and class").
9. Richard White and William Cronon, *Remembering Ahanagran: A History of Stories* (Seattle: University of Washington Press, 2003).
10. Some other expressions of this include Katie Holmes and Heather Goodall, eds., *Telling Environmental Histories: Intersections of Memory, Narrative and Environment* (New York: Palgrave Macmillan, 2017); Debbie Lee and Kathryn Newfont, eds., *The Land Speaks: New Voices at the Intersection of Oral and Environmental History* (New York: Oxford University Press, 2017); and Brian Williams, "The Challenge of Oral History to Environmental History," *Environment and History* (2018): 1–25.
11. Dipesh Chakrabarty, *Provincializing Europe* (Princeton, NJ: Princeton University Press, 2000); for another earlier influential essay along similar lines: Gayatri Chakravorty Spivak, "Can the Subaltern Speak?" in *Marxism and the Interpretation of Culture*, ed. C. Nelson and L. Grossberg (Basingstoke: Macmillan Education, 1988), 271–313. http://abahlali.org/files/Can_the_subaltern_speak.pdf.
12. Harold K. Steen and Susan R. Schrepfer, "Why Oral History?" *Forest History Newsletter* 16, no. 3 (1972): 4–5. https://doi.org/10.2307/4004282.
13. Matthew S. Champion, "History of Temporalities: An Introduction," *Past & Present* 243, no. 1 (May 2019): 247–54. https://doi.org/10.1093/pastj/gtz007; Matthew S. Champion, "A Fuller History of Temporalities," *Past & Present* 243, no. 1 (May 1, 2019): 255–66. https://doi.org/10.1093/pastj/gtz008; A. R. P. Fryxell, "Time and the Modern: Current Trends in The History of Modern Temporalities," *Past & Present* 243, no. 1 (May 2019): 285–98. https://doi.org/10.1093/pastj/gtz012; Vanessa Ogle, "Time, Temporality and the History of Capitalism," *Past & Present* 243, no. 1 (May 2019): 312–27. https://doi.org/10.1093/pastj/gtz014.
14. Marco Armiero and Richard Tucker, eds., *Environmental History of Modern Migrations* (London; New York: Routledge, 2017).
15. William Cronon, "A Place for Stories: Nature, History, and Narrative." *Journal of American History* 78, no. 4 (March 1992): 1347–76.
16. Ted Steinberg, "Down, Down, Down, No More," *Journal of the Early Republic* 24, no. 2 (Summer 2004): 260–66; Brian Donahue, *The Great Meadow: Farmers and the Land in Colonial Concord* (New Haven, CT: Yale University Press, 2004); Carolyn Merchant, *Reinventing Eden: The Fate of Nature in Western Culture* (New York: Routledge, 2003); Timothy J. LeCain, *Mass Destruction: The Men and Giant Mines That Wired America and Scarred the Planet* (New Brunswick, NJ: Rutgers University Press, 2009); Christopher Sellers, *Crabgrass Crucible: Suburban Nature and the Rise of Environmentalism in Twentieth-Century America* (Chapel Hill: University of North Carolina Press, 2012); Victoria Penziner Hightower, "Pearls and the Southern Persian/Arabian Gulf: A Lesson in Sustainability," *Environmental History* 18, no. 1 (January 1, 2013): 44–59. https://doi.org/10.1093/envhis/ems113.
17. Holmes and Goodall, *Telling Environmental Histories*, 340.
18. Mark Fiege, *The Republic of Nature* (University of Washington, 2012); Andrew C. Isenberg, "Introduction; A New Environmental History," in *The Oxford Handbook of*

Environmental History, ed. Andrew C. Isenberg (New York: Oxford University Press, 2014), 1–20.

19. Ian Hacking, *Historical Ontology* (Cambridge, MA: Harvard University Press, 2004); Ursula Klein and Wolfgang Lefevre, *Materials in Eighteenth-Century Science: A Historical Ontology* (Cambridge, MA: MIT Press, 2007); Christopher Sellers, "The Artificial Nature of Fluoridated Water: Between Nations, Knowledges, and Material Flows," in *Osiris—Landscapes of Exposure; Knowledge and Illness in Modern Environments*, ed. Gregg Mitman, Michelle Murphy, and Christopher Sellers, vol. 19, (Chicago: University of Chicago Press, 2004), 182–200; David C. Posthumus, "All My Relatives: Exploring Nineteenth-Century Lakota Ontology and Belief," *Ethnohistory* 64, no. 3 (July 2017): 379–400 ("alter worlds"). https://doi.org/10.1215/00141801-3870627; Valentina Bonifacio, "Meeting the Generals: A Political Ontology Analysis of the Paraguayan Maskoy Struggle for Land," *Anthropologica* 55, no. 2 (September 2013): 385–97; Colin Hay, "Neither Real nor Fictitious but 'as If Real'? A Political Ontology of the State," *British Journal of Sociology* 65, no. 3 (2014): 459–80, https://doi.org/10.1111/1468-4446.12082; Greg Anderson, "Retrieving the Lost Worlds of the Past: The Case for an Ontological Turn," *American Historical Review* 120, no. 3 (June 2015): 787–810. https://doi.org/10.1093/ahr/120.3.787.
20. Lee and Newfont, *The Land Speaks*; Aldo Leopold, *Aldo Leopold: A Sand County Almanac & Other Writings on Conservation and Ecology*, ed. Curt Meine, Reprint edition (New York: Library of America, 2013).
21. Janet Conway and Jakeet Singh, "Radical Democracy in Global Perspective: Notes from the Pluriverse," *Third World Quarterly* 32, no. 4 (2011): 702 ("cosmopolitan"); 701 ("multiple ontologies").

References

Books

Abbott, Dina, and Gordon Wilson. *The Lived Experience of Climate Change: Knowledge, Science and Public Action*. New York: Springer, 2015.

Alexievich, Svetlana. *Voices from Chernobyl*. Normal, Illinois: Dalkey Archive Press, 2005.

Anderson, Deb. *Endurance: Australian Stories of Drought*. Collingwood, Australia: CSIRO Publishing, 2014.

Aso, Michitake. *Rubber and the Making of Vietnam: An Ecological History, 1897–1975*. Chapel Hill: University of North Carolina Press, 2018.

Beardy, Flora, and Robert Coutts, eds., *Voices from Hudson Bay: Cree Stories from York Factory*. Montreal; Kingston: McGill-Queen's University Press, 1996.

Benac, David. *Conflict in the Ozarks: Hill Folk, Industrialists, and Government in Missouri's Courtois Hills*. Kirksville, MO: Truman State University Press, 2010.

Berkes, Fikret. *Sacred Ecology: Traditional Ecological Knowledge and Resource Management*. London: Routledge, 1999.

Blong, Russell J. *The Time of Darkness: Local Legends and Volcanic Reality in Papua New Guinea*. Canberra: Australian National University Press, 1982.

Blum, Elizabeth. *Love Canal Revisited: Race, Class, and Gender in Environmental Activism*. Lawrence: University of Kansas Press, 2004.

Bristow, Tom, and Thomas H. Ford, eds. *A Cultural History of Climate Change*. Routledge Environmental Humanities Series. New York: Routledge, 2016.

Brown, Malcolm, and Stephen Forbes. *Whisperings of the Wannon: An Oral and Ecological History of the Wannon River in South West Victoria*. Hamilton, Victoria, Australia: Glenelg-Hopkins Catchment Management Authority, 2002.

Brugge, Doug, Timothy Benally, and Esther Yazzie-Lewis. *The Navajo People and Uranium Mining*. Albuquerque: University of New Mexico Press, 2006.

Cave, Mark, and Stephen Sloan, eds. *Listening on the Edge: Oral History in the Aftermath of Crisis*. New York: Oxford University Press, 2014.

Chiarappa, Michael J., and Kristin M. Szylvian. *Fish for All: An Oral History of Multiple Claims and Divided Sentiment on Lake Michigan*. East Lansing: Michigan State University Press, 2003.

Conte, Christopher Allan. *Highland Sanctuary: Environmental History in Tanzania's Usambara Mountains*. Athens: Ohio University Press, 2004.

Convery, Ian, Gerard Corsane, and Peter Davis. *Displaced Heritage: Responses to Disaster, Trauma, and Loss*. Woodbridge, Suffolk, UK: The Boydell Press, 2014.

Crow, Peter. *Do, Die, or Get Along: A Tale of Two Appalachian Towns*. Athens: University of Georgia Press, 2007.

Cruikshank, Julie. *Do Glaciers Listen—Local Knowledge, Colonial Encounters and Social Imagination*. Vancouver: University of British Columbia Press, 2005.

Dovers, Stephen. *Environmental History and Policy: Still Settling Australia.* Melbourne: Oxford University Press, 2000.

Evans, George Ewart. *The Pattern under the Plough: Aspects of the Folk-life of East Anglia.* London: Faber, 1966.

Fairhead, James, and Melissa Leach. *Misreading the African Landscape: Society and Ecology in a Forest-savanna Mosaic.* Cambridge; New York: Cambridge University Press, 1996.

Ferguson, Cody. *This Is Our Land: Grassroots Environmentalism in the Late Twentieth Century.* New Brunswick, NJ: Rutgers University Press, 2015.

Frankland, Peggy, and Susan Tucker. *Women Pioneers of the Louisiana Environmental Movement.* Jackson: University Press of Mississippi, 2013.

Friederici, Peter, ed. *What Has Passed and What Remains: Oral Histories of Northern Arizona's Changing Landscapes.* Tucson: University of Arizona Press, 2010.

Gaggio, Dario. *The Shaping of Tuscany: Landscape and Society between Tradition and Modernity.* Cambridge, UK: Cambridge University Press, 2017.

Georgette, Susan. *In the Rough Land to the South: An Oral History of the Lives and Events at Big Creek, Big Sur, California.* Santa Cruz: Environmental Field Program, University of California, Santa Cruz, 1981.

Gibbs, Lois Marie, and Murray Levine. *Love Canal: My Story.* Albany: State University of New York Press, 1982.

Hay, Iain, ed. *Qualitative Research Methods in Human Geography.* Fourth edition. Oxford: Oxford University Press, 2016.

Helander-Renvall, Elina, and Tero Mustonen. *Snowscapes, Dreamscapes: Snowchange Book on Community Voices of Change.* Vassa, Finland: Tampere Polytechnic, 2004.

Holmes, Katie, and Heather Goodall. *Telling Environmental Histories: Intersections of Memory, Narrative and Environment.* Cham, Switzerland: Palgrave Macmillan, 2017.

House, Silas, and Jason Howard. *Something's Rising: Appalachians Fighting Mountaintop Removal.* Lexington: University Press of Kentucky, 2009.

Hurley, Andrew. *Environmental Inequalities: Class, Race, and Industrial Pollution in Gary, Indiana, 1945–1980.* Chapel Hill: The University of North Carolina Press, 2009.

Hussey, Stephen, and Paul Thompson. *The Roots of Environmental Consciousness.* London: Routledge, 2005.

Janus, Edward. *Creating Dairyland: How Caring for Cows Saved Our Soil, Created Our Landscape, Brought Prosperity to Our State, and Still Shapes Our Way of Life in Wisconsin.* Madison: Wisconsin Historical Society Press, 2011.

Johannes, Robert Earle. *Words of the Lagoon: Fishing and Marine Lore in the Palau District of Micronesia.* Berkeley: University of California Press, 1981.

Johnson, Sarah. *Indigenous Knowledge.* Knapwell, Winwick, UK: White Horse Press, 2012.

Keeling, Arn, and John Sandlos. *Mining and Communities in Northern Canada: History, Politics, and Memory.* Calgary: University of Calgary Press, 2015.

Kimura, Aya Hirata. *Radiation Brain Moms and Citizen Scientists: The Gender Politics of Food Contamination after Fukushima.* Durham, NC: Duke University Press, 2016.

Kohn, Eduardo. *How Forests Think: Toward an Anthropology beyond the Human.* Berkeley: University of California Press, 2013.

Kolvet, Renee Corona, and Victoria Ford. *The Civilian Conservation Corps in Nevada: From Boys to Men.* Reno: University of Nevada Press, 2006.

Lee, Debbie, and Kathryn Newfont, eds. *The Land Speaks: New Voices at the Intersection of Oral and Environmental History.* New York: Oxford University Press, 2017.

Leech, Brian James. *The City That Ate Itself: Butte, Montana and Its Expanding Berkeley Pit*. Reno: University of Nevada Press, 2018.

Luby, Brittany. *Dammed: The Politics of Loss and Survival in Anishinaabe Territory*. Winnipeg: University of Manitoba Press, 2020.

MacDonald, John. *The Arctic Sky: Inuit Astronomy, Star Lore, and Legend*. Toronto: Royal Ontario Museum, 1998.

Marsh, George Perkins. *Man and Nature: Or, Physical Geography as Modified by Human Action*. Cambridge, Massachusetts: Belknap Press of Harvard University Press, 1965 (Originally published in 1864 by Scribner, New York).

McDonald, Miriam Anne, Lucassie Arragutainaq, and Zacharassie Novalinga. *Voices from the Bay: Traditional Ecological Knowledge of Inuit and Cree in the Hudson Bay Bioregion*. Ottawa: Canadian Arctic Resources Committee, and Environmental Committee of Sanikiluaq, 1997.

McLeod, Maria. *Washington State Department of Ecology Historically Speaking: An Oral History in Celebration of the First 35 Years, 1970–2005*. Olympia: Washington State Department of Ecology, 2005.

Menzies, Charles R., ed. *Traditional Ecological Knowledge and Natural Resource Management*. Lincoln: University of Nebraska Press, 2006.

Mukherjee, Suroopa. *Dancing Bodies, Written Texts, and Oral Testimonials of Women in the Wake of an Industrial Disaster*. Basingstoke, UK: Palgrave Macmillan, 2010.

Oiwa, Keibo, Masato Ogata, and Karen Colligan-Taylor. *Rowing the Eternal Sea: The Story of a Minamata Fisherman*. Lanham, MD: Rowman & Littlefield, 2001.

Perez, Karni R. *Fishing for Gold: The Story of Alabama's Catfish Industry*. Tuscaloosa: University of Alabama Press, 2006.

Peterson, Anna L. *Being Animal: Beasts and Boundaries in Nature Ethics*. New York: Columbia University Press, 2013.

Riney-Kehrberg, Pamela. *Rooted in Dust: Surviving Drought and Depression in Southwestern Kansas*. Lawrence: University Press of Kansas, 1994.

Roberts, Jane, and Geoff R. Sainty. *Oral History as a Tool in Historical Ecology: Lachlan River as a Case Study*. New South Wales, Australia: CSIRO Publishing, 1997.

Rogers, Peter. *Water Crisis: Myth or Reality*. London: Taylor and Francis, 2006.

Rothschild, Rachel Emma. *Poisonous Skies: Acid Rain and the Globalization of Pollution*. Chicago: University of Chicago Press, 2019.

Russell, Emily W. B. *People and the Land through Time: Linking Ecology and History*. New Haven, CT: Yale University Press, 1998.

Ryden, Kent C. *Mapping the Invisible Landscape Folklore, Writing, and the Sense of Place*. Iowa City: University of Iowa Press, 1993.

Sanders, Angela. *Oral Histories Documenting Changes in Wheat Belt Wetlands*. Como: Western Australia Department of Conservation and Land Management, 1991.

Schama, Simon. *Landscape and Memory*. First American edition. New York: A. A. Knopf; Distributed by Random House, 1995.

Scott, Anthony. *The Ecology of the Tuggerah Lakes: An Oral History*. Canberra, Australia: CSIRO Publishing, 1999.

Shetler, Jan Bender. *Imagining Serengeti: A History of Landscape Memory in Tanzania from Earliest Times to the Present*. Athens: Ohio University Press, 2007.

Shiva, Vandana. *Water Wars: Privatization, Pollution and Profit*. Berkeley, CA: North Atlantic Books, 2016.

Shoemaker, Nancy, ed. *Living with Whales: Documents and Oral Histories of Native New England Whaling History*. Amherst: University of Massachusetts Press, 2014.

Skaria, Ajay. *Hybrid Histories: Forests, Frontiers, and Wildness in Western India*. Delhi; New York: Oxford University Press, 1999.

Slongwhite, Dale Finley. *Fed Up: The High Costs of Cheap Food*. Gainesville: University Press of Florida, 2014.

Tiffen, Mary, Michael Mortimore, and Francis Gichuki. *More People, Less Erosion: Environmental Recovery in Kenya*. London: John Wiley & Sons Ltd, 1994.

Tittensor, Ruth. *From Peat Bog to Conifer Forest: An Oral History of Whitelee, Its Community and Landscape*. Chichester, UK: Packard Publishing, 2009.

Todd, David A., and David Weisman, eds. *The Texas Legacy Project: Stories of Courage and Conservation*. College Station: Texas A&M University Press, 2010.

Tompkins, Andrew S. *Better Active Than Radioactive!: Anti-Nuclear Protest in 1970s France and West Germany*. Oxford: Oxford University Press, 2016.

Ulturgasheva, Olga. *Narrating the Future in Siberia: Childhood, Adolescence and Autobiography among the Eveny*. New York: Berghahn Books, 2012.

Urton, Gary. *At the Crossroads of the Earth and the Sky: An Andean Cosmology*. First edition. Austin: University of Texas Press, 1981.

Valenčius, Conevery Bolton. *The Lost History of the New Madrid Earthquakes*. Chicago: University of Chicago Press, 2015.

Weisiger, Marsha L. *Dreaming of Sheep in Navajo Country*. Seattle: University of Washington Press, 2009.

Zhou, Xun. *Forgotten Voices of Mao's Great Famine, 1958–1962*. New Haven, CT: Yale University Press, 2013.

Articles and Book Chapters

Alexis-Martin, Becky. "The Chernobyl Necklace: The Psychosocial Experiences of Female Radiation Emergency Survivors." *Belgeo* 1 (2015): 1–12.

Anderson, Deb. "Voice of Endurance: Climate and the Power of Oral History." In *A Cultural History of Climate Change*, edited by Tom Bristow and Thomas H. Ford, 17–37. New York: Routledge, 2016.

Arce-Nazario, Javier A. "Landscape Images in Amazonian Narrative: The Role of Oral History in Environmental Research." *Conservation and Society* 5, no. 1 (2007): 115–33.

Arias, Ana Yolanda Zúñiga. "Ríos, cultivos e inundaciones: historia reciente del cantón de Corredores, Costa Rica, 1950–2010." *Ambiente y Desarrollo* 18, no. 34 (2014): 77–92.

Barca, Stefania, and Ana Delicado. "Anti-Nuclear Mobilisation and Environmentalism in Europe: A View from Portugal (1976–1986)." *Environment and History* 22, no. 4 (November 2016): 497–520. https://doi.org/10.3197/096734016X14727286515736.

Bendix, Regina. "Reflections on Earthquake Narratives." *Western Folklore* 49, no. 4 (1990): 331–47.

Berkes, Fikret, Carl Folke, and Madhav Gadgil. "Traditional Ecological Knowledge, Biodiversity, Resilience and Sustainability." In *Biodiversity Conversations: Problems and Policies*, edited by Charles A. Perrings, Karl-Göran Mäler, Carl Folke, Crawford Stanley Holling, and Bengt-Owe Jansson, 281–99. Dordrecht, Netherlands: Springer Dordrecht, 1995. https:// doi.org/10.1007/ 978-94-011-0277-3_15.

Bhandari, Roshan Bhakta, Norio Okada, and J. David Knottnerus. "Urban Ritual Events and Coping with Disaster Risk: A Case Study of Lalitpur, Nepal." *Journal of Applied Social Science* 5, no. 2 (2011): 13–32.

Bingley, Amanda. "Health, People and Forests." In *Making Sense of Place: Multidisciplinary Perspectives*, edited by Ian Convery, Gerard Corsane, and Peter Davis, 107-16. Woodbridge, Suffolk, UK: Boydell & Brewer, 2012.

Bird, Kristin E., Wallace J. Nichols, and Charles R. Tambiah. "The Value of Local Knowledge in Sea Turtle Conversation: A Case from Baja California, Mexico." Presented at Putting Fishers' Knowledge to Work: Conference Proceedings August 27–30, 2001. University of British Columbia Fisheries Centre Research Reports 11 (January 2003): 178–83.

Brummond, Janice. "Liquidators, Chornobylets and Masonic Ecologists: Ukrainian Environmental Identities." *Oral History* 28, no. 1 (2000): 52–62.

Byram, R. Scott. "Tectonic History and Cultural Memory: Catastrophe and Restoration on the Oregon Coast." *Oregon Historical Quarterly* 108, no. 2 (2007): 167–80.

Calheiros, D. F., A. F. Seidl, and C. J. A. Ferreira. "Participatory Research Methods in Environmental Science: Local and Scientific Knowledge of a Limnological Phenomenon in the Pantanal Wetland of Brazil." *Journal of Applied Ecology* 37, no. 4 (2000): 684–96.

Cary, David, Jr. "A Cautionary Tale of Environmental Management: Malaria, Water Management, and Land Reclamation in Twentieth-Century Guatemala." *Environmental History* 26, no. 3 (July 2021): 555–80. https://doi.org/10.1093/envhis/emab023.

Carter, Ian. "Oral History and Agrarian History: The North East." *Oral History* 2, no 1 (1974): 34–44.

Clark, Brett, John Bellamy Foster, and George Perkins Marsh. "George Perkins Marsh and the Transformation of Earth: An Introduction to Marsh's Man and Nature / Excerpts from Man and Nature." *Organization & Environment* 15, no. 2 (2002): 164–77.

Colasurdo, Christine. "Oregon Voices: Recollecting a Landscape: Oral Histories of Spirit Lake and Mount St. Helens." *Oregon Historical Quarterly* 101, no. 3 (2000): 364–75.

Cole, Tim. "'Nature Was Helping Us': Forests, Trees, and Environmental Histories of the Holocaust." *Environmental History* 19, no. 4 (October 2014): 665–86. https://doi.org/10.1093/envhis/emu068.

Compton, Kimberly S. "Pro-Environmental Behaviors among Black Environmentalists: A Critical Race Perspective." PhD diss. Virginia Commonwealth University, 2021.

Conte, Christopher. "Africa's Mountains: Collecting and Interpreting the Past." *RCC Perspectives*, no. 4 (2014): 55–64.

Cooper, Timothy, and Anna Green. "The *Torrey Canyon* Disaster, Everyday Life, and the 'Greening' of Britain." *Environmental History* 22, no. 1 (January 2017): 101–26.

Cregeen, E. R. "Oral Tradition and Agrarian History in the West Highlands." *Oral History* 2, no. 1 (Spring 1974): 15–33.

Cronon, William. "A Place for Stories: Nature, History, and Narrative." *Journal of American History* 78, no. 4 (March 1, 1992): 1347–76.

Crook, Darren S. "The Historical Impacts of Hydroelectric Power Development on Traditional Mountain Irrigation in the Valais, Switzerland." *Mountain Research and Development* 21, no. 1 (February 2001): 46–53.

Cross, Nigel, and Rhiannon Barker. "The Sahel Oral History Project." In *The Oral History Reader*, edited by Robert Perks and Alistair Thomson, 246–57. First Edition. London: Routledge, 1997.

Cruikshank, Julie. "Glaciers and Climate Change: Perspectives from Oral Tradition." *Arctic* 54, no. 4 (December 2001): 377–93.

Cruikshank, Julie. "Legend and Landscape: Convergence of Oral and Scientific Traditions in the Yukon Territory." *Arctic Anthropology* 18, no. 2 (1981): 67–93.

_____. "Oral and Written Interpretations of the Past." In *Reading Voices: Dän Dhá Ts'edenintthę̂: Oral and Written Interpretations of the Yukon's Past* by Julie Cruikshank, 11–17. Vancouver: Douglas & McIntyre, 1991.

Dahlberg, Annika C., and Piers M. Blaikie. "Changes in Landscape or in Interpretation? Reflections Based on the Environmental and Socio-Economic History of a Village in NE Botswana." *Environment and History* 5, no. 2 (June 1999): 127–74.

Daley, Ben, and Peter Griggs. "'Loved to Death': Coral Collecting in the Great Barrier Reef, Australia, 1770–1970." *Environment and History* 14, no. 1 (2008): 89–119. https://doi:10.3197/096734008X271878.

_____. "Mining the Reefs and Cays: Coral, Guano and Rock Phosphate Extraction in the Great Barrier Reef, Australia, 1844–1940." *Environment and History* 12, no. 4 (2006): 395–433.

_____, and Helene Marsh. "Exploiting Marine Wildlife in Queensland: The Commercial Dugong and Marine Turtle Fisheries, 1847–1969." *Australian Economic History Review* 48, no. 3 (2008): 227–65.

_____. "Reconstructing Reefs: Qualitative Research and the Environmental History of the Great Barrier Reef, Australia." *Qualitative Research: QR* 8, no. 5 (2008): 584–615.

Davis, J., and B. Finlayson. "The Role of Historical Research in Stream Rehabilitation: A Case Study from Central Victoria." *The Challenge of Rehabilitating Australia's Streams* 1 (1999): 199–204.

Dudley, Walter C., James Goff, Catherine Chagué, and Jeanne Johnston. "Capturing the Next Generation of Cultural Memories – The Process of Interviewing Tsunami Survivors." *Science of Tsunami Hazards* 28, no. 3 (January 1, 2009): 154–70.

Egoz, Shelley. "Clean and Green but Messy: The Contested Landscape of New Zealand's Organic Farms." *Oral History* 28, no. 1 (2000): 63–74.

Endres, Danielle. "Environmental Oral History." *Environmental Communication* 5, no. 4 (December 1, 2011): 485–98.

Finlayson, Brian L., and Sandra O. Brizga. "The Oral Tradition, Environmental Change and River Basin Management: Case Studies from Queensland and Victoria." *Australian Geographical Studies* 33, no. 2 (October 1995): 180–92. https://doi:10.1111/j.1467-8470.1995.tb00693.x.

Fogerty, James E. "Oral History: A Guide to Its Creation and Use." In *The Historical Ecology Handbook: A Restorationist's Guide to Reference Ecosystems*, edited by Evelyn A. Howell and Dave. Egan, 101–20. Washington, DC: Island Press, 2001.

Gant, Robert. "Oral History and Local Meteorology." *Oral History* 14, no. 2 (1986): 67–9.

Gaynor, Andrea, and Joy McCann. "'I've Had Dolphins . . . Looking for Abalone for Me': Oral History and the Subjectivities of Marine Engagement." *Oral History Review* 44, no. 2 (Summer/Fall 2017): 260–77.

Gearheard, Shari, Matthew Pocernich, Ronald Stewart, Joelie Sanguya, and Henry P. Huntington. "Linking Inuit Knowledge and Meteorological Station Observations to Understand Changing Wind Patterns at Clyde River, Nunavut." *Climatic Change* 100, no. 2 (2009): 267–94.

Gimmi, Urs, and Matthias Burgi. "Using Oral History and Forest Management Plans to Reconstruct Traditional Non-Timber Forest Uses in the Swiss Rhone Valley since the Late Nineteenth Century." *Environment and History* 13, no. 2 (2007): 211–46.

Goodall, Heather. "Riding the Tide: Indigenous Knowledge, History and Water in a Changing Australia." *Environment and History* 14, no. 3 (2008): 355–84.

Green, Anna, and Timothy Cooper. "Community and Exclusion: The *Torrey Canyon* Disaster of 1967." *Journal of Social History* 48, no. 4 (Summer 2015): 892–909.

Hague, Euan. "Landscapes of Memory: Oral History and the Environment." *Journal of Historical Geography* 26, no. 1 (January 1, 2000): 115–17.

Hanson, Anne-Marie S. "Shoes in the Seaweed and Bottles on the Beach: Global Garbage and Women's Oral Histories of Socio-Environmental Change in Coastal Yucatán." In *A Political Ecology of Women, Water and Global Environmental Change*, edited by Stephanie Buechler and Anne-Marie S. Hanson, 165–84. London: Routledge, 2015.

Hanson, Anne-Marie S. "Women's Ecological Oral Histories of Recycling and Development in Coastal Yucatán." *Gender, Place & Culture* 23, no. 4 (2016): 467–83. https://doi.org/10.1080/0966369X.2015.1013445.

Harvey, David. "Country Stories: The Use of Oral Histories of the Countryside to Challenge the Sciences of the Past and Future." *Interdisciplinary Science Reviews* 30, no. 1 (March 1, 2005): 19–32.

Holmes, Andrew, and Margaret Pilkington. "Storytelling, Floods, Wildflowers and Wash Lands: Oral History in the River Ouse Project." *Oral History* 39, no. 2 (2011): 83–94.

Hubbard, Blair, Heather Huyck, and David Nathanson. *Collecting, Using and Preserving Oral History in the National Park Service*. Harpers Ferry: US Department of the Interior, 1984.

Huntington, Henry P. "Using Traditional Ecological Knowledge in Science: Methods and Applications." *Ecological Applications* 10, no. 5 (2000): 1270–74.

Isaacman, Allen, and Chris Sneddon. "Toward a Social and Environmental History of the Building of Cahora Bassa Dam." *Journal of Southern African Studies* 26, no. 4 (2000): 597–632.

Jacobs, Nancy. "Grasslands and Thickets: Bush Encroachment and Herding in the Kalahari Thornveld." *Environment and History* 6, no. 3 (2000): 289–316.

Janif, Shaiza Z., Patrick D. Nunn, Paul Geraghty, William Aalbersberg, Frank R. Thomas, and Mereoni Camailakeba. "Value of Traditional Oral Narratives in Building Climate-Change Resilience: Insights from Rural Communities in Fiji." *Ecology and Society* 21, no. 2 (2016): 7–16.

Johnston, Jeanne Branch. "Personal Accounts from Survivors of the Hilo Tsunamis of 1946 and 1960: Toward a Disaster Communication Model." MA Thesis, University of Hawaii, 2003.

Johnston, Jeanne Branch. *Tsunamis in Maui County: Oral Histories*. Honolulu: Center for Oral History, Social Science Research Institute, University of Hawaii at Manoa, 2003.

Jung, Christian A., Peter D. Dwyer, Monica Minnegal, and Stephen E. Swearer. "Perceptions of Environmental Change over More Than Six Decades in Two Groups of People Interacting with the Environment of Port Phillip Bay, Australia." *Ocean & Coastal Management* 54, no. 1 (January 2011): 93–9.

King, Darren N., and James R. Goff. "Benefitting from Differences in Knowledge, Practice and Belief: Maori Oral Traditions and Natural Hazards Science." *Natural Hazards and Earth System Sciences; Katlenburg-Lindau* 10, no. 9 (September 2010): 1927–40. https://doi.org/10.5194/nhess-10-1927-2010.

King, Darren N., Wendy S. Shaw, Peter N. Meihana, and James R. Goff. "Māori Oral Histories and the Impact of Tsunamis in Aotearoa-New Zealand." *Natural Hazards and Earth System Sciences; Katlenburg-Lindau* 18, no. 3 (2018): 907–19.

Klein, Jørgen, David J. Nash, Kathleen Pribyl, Georgina H. Endfield, and Matthew J. Hannaford. "Climate, Conflict and Society: Changing Responses to Weather Extremes in Nineteenth Century Zululand." *Environment and History* 24, no. 3 (August 2018): 377–401. https://doi:10.3197/096734018X15137949591963.

Kuszmar, Kate. "From Boat to Throat: How Oral Histories Immerse Students in Eco literacy and Community Building." *Oral History Review* 41, no. 2 (2014): 325–40.

Lane, Ruth. "Oral Histories and Scientific Knowledge in Understanding Environmental Change: A Case Study in the Tumut Region, NSW." *Australian Geographical Studies* 35, no. 2 (1997): 195–205.

Lauer, Matthew. "Oral Traditions or Situated Practices? Understanding How Indigenous Communities Respond to Environmental Disasters." *Human Organization* 71, no. 2 (2012): 176–87.

Leal, Claudia. "Conservation Memories: Vicissitudes of a Biodiversity Conservation Project in the Rainforests of Colombia, 1992–1998." *Environmental History* 20, no. 3 (2015): 368–95. https://doi.org/10.1093/envhis/emv051.

Leddy, Lianne. "Interviewing Nookomis and Other Reflections: The Promise of Community Collaboration." *Oral History Forumd' histoire orale 30*, Special Issue "Talking Green: Oral History and Environmental History" (2010): 1–18. https://www.oralhistoryforum.ca/index.php/ohf/article/download/386/457/ Lee, Debbie. "Listening to the Land: The Selway-Bitterroot Wilderness as Oral History." *The Oral History Review* 37, no. 2 (Summer/Fall 2010): 235–48.

Lemelin, Raynald Harvey, Martha Dowsley, Brian Walmark, Franz Siebel, Louis Bird, George Hunter, Tommy Myles, Maurice Mack, Matthew Gull, and Matthew Kakekaspan. "Wabusk of the Omushkegouk: Cree-Polar Bear (Ursus Maritimus) Interactions in Northern Ontario." *Human Ecology: An Interdisciplinary Journal* 38, no. 6 (November 23, 2010): 803–15.

Leonard, Sonia, and Jonathan Nott. "Integrating Geomorphological Assessments and Oral Histories in Catchment Management: Implications for the Mulgrave River, Northeastern Australia." *Geographical Research* 53, no. 2 (2015): 160–74.

Leonard, Sonia, Meg Parsons, Knut Olawsky, and Frances Kofod. "The Role of Culture and Traditional Knowledge in Climate Change Adaptation: Insights from East Kimberley, Australia." *Global Environmental Change* 23, no. 3 (June 1, 2013): 623–32.

Liu, Pingyang, Juan Moreno, Peiying Song, Elona Hoover, and Marie Harder. "The Use of Oral Histories to Identify Criteria for Future Scenarios of Sustainable Farming in the South Yangtze River, China." *Sustainability* 8, no. 9: 859 (2016): 1–24. https://doi.org/10.3390/su8090859.

Longley, Hereward. "Conflicting Interests: Development Politics and the Environmental Regulation of the Alberta Oil Sands Industry, 1970–1980." *Environment & History* 27, no. 1 (February 2021): 97–125. https://doi.org/10.3197/096734019X15463432086919.

Lykke, A. M. "Local Perceptions of Vegetation Change and Priorities for Conservation of Woody-Savanna Vegetation in Senegal." *Journal of Environmental Management* 59, no. 2 (June 1, 2000): 107–20. https://doi.org/10.1006/jema.2000.0336.

Mallory, Mark L., H. Grant Gilchrist, Alain J. Fontaine, and Jason A. Akearok. "Local Ecological Knowledge of Ivory Gull Declines in Arctic Canada." *Arctic* 56, no. 3 (January 2003): 293–98.

McDonnell, Janet A. "Documenting Cultural and Historical Memory: Oral History in the National Park Service." *Oral History Review* 30, no. 2 (2003): 99–109.

Milligan, Christine, and Amanda Bingley. "Restorative Places or Scary Spaces? The Impact of Woodland on the Mental Well-Being of Young Adults." *Health & Place* 13, no. 4 (December 1, 2007): 799–811.

Mustonen, Tero. "Oral Histories as a Baseline of Landscape Restoration – Co-Management and Watershed Knowledge in Jukajoki River." *Fennia: International Journal of Geography* 191, no. 2 (August 1, 2013): 76–91.

Neufeld, David. "Parks Canada, The Commemoration of Canada, and Northern Aboriginal Oral History." In *Oral History and Public Memories*, edited by Paula Hamilton and Linda Shopes, 7–30. Philadelphia: Temple University Press, 2008.

Newman, Rich. "Making Environmental Politics: Women and Love Canal Activism." *Women's Studies Quarterly* 29, no. 1/2 (2001): 65–84.

Nichols, Theresa, Fikret Berkes, Dyanna Jolly, Norman B. Snow, and The Community of Sachs Harbour. "Climate Change and Sea Ice: Local Observations from the Canadian Western Arctic." *Arctic* 57, no. 1 (March 2004): 68–79.

Olafsdottir, Gunnthora, Paul Cloke, and Claus Vögele. "Place, Green Exercise and Stress: An Exploration of Lived Experience and Restorative Effects." *Health & Place* 46 (July 1, 2017): 358–65.

Olsson, Per, and Carl Folke. "Local Ecological Knowledge and Institutional Dynamics for Ecosystem Management: A Study of Lake Racken Watershed, Sweden." *Ecosystems* 4, no. 2 (March 2001): 85–104.

Oosthoek, Jan. "The Role of Oral History in Environmental History." Paper presented at the annual meeting of the European Association for Environmental History, UK Branch. Open University, Milton Keynes, May 21, 2004. https://www.eh-resources.org/role-of-oral-history-in-environmental-history/.

Perkins, Tracy E. "Women's Pathways Into Activism: Rethinking the Women's Environmental Justice Narrative in California's San Joaquin Valley." *Organization & Environment* 25, no. 1 (2012): 76–94.

Phelan, Michael J. "Sciaenid Aggregations in Northern Australia: An Example of Successful Outcomes through Collaborative Research." Presented at Putting Fishers' Knowledge to Work: Conference Proceedings August 27–30, 2001. University of British Columbia Fisheries Centre Research Reports 11 (January 2003): 100–09. http://dx.doi.org/10.14288/1.0074793.

Piper, Liza. "Subterranean Bodies: Mining the Large Lakes of North-West Canada, 1921–1960." *Environment and History* 13, no. 2 (2007): 155–86.

Riedlinger, Dyanna, and Fikret Berkes. "Contributions of Traditional Knowledge to Understanding Climate Change in the Canadian Arctic." *Polar Record* 37, no. 203 (2001): 315–28. https://doi.org/10.1017/S0032247400017058.

Riley, Mark. "Ask the Fellows Who Cut the Hay: Farm Practices, Oral History and Nature Conservation." *Oral History* 32, no. 2 (2004): 45–53.

Riley, Mark, and David Harvey. "Landscape Archaeology, Heritage and the Community in Devon: An Oral History Approach." *International Journal of Heritage Studies* 11, no. 4 (January 1, 2005): 269–88. https://doi.org/10.1080/13527250500235567

_____. "Oral Histories, Farm Practice and Uncovering Meaning in the Countryside." *Social & Cultural Geography* 8, no. 3 (June 2007): 391–415. https://doi.org/10.1080/14649360701488823.

Roberts, Jane, and Geoff Sainty. "Oral History, Ecological Knowledge, and River Management." In *Environmental History and Policy: Still Settling Australia*, edited by Stephen Dovers, 118–44. Melbourne: Oxford University Press, 2000.

Robertson, Hugh A., and Tara K. McGee. "Applying Local Knowledge: The Contribution of Oral History to Wetland Rehabilitation at Kanyapella Basin, Australia." *Journal of*

Environmental Management 69, no. 3 (November 2003): 275–87. https://doi.org/10.1016/S0301-4797(03)00155-5.

Robertson, Margaret, Pam Nichols, Pierre Horwitz, Keith Bradby, and David MacKintosh. "Environmental Narratives and the Need for Multiple Perspectives to Restore Degraded Landscapes in Australia." *Ecosystem Health* 6, no. 2 (June 2000): 119–33. https://doi.org/10.1046/j.1526-0992.2000.00013.x

Russell, Roly, Anne D. Guerry, Patricia Balvanera, Rachelle K. Gould, Xavier Basurto, Kai M. A. Chan, Sarah Klain, Jordan Levine, and Jordan Tam. "Humans and Nature: How Knowing and Experiencing Nature Affect Well-Being." *Annual Review of Environment and Resources* 38, no. 1 (October 17, 2013): 473–502.

Sáenz-Arroyo, Andrea, and Daniel Revollo-Fernández. "Local Ecological Knowledge Concurs with Fishing Statistics: An Example from the Abalone Fishery in Baja California, Mexico." *Marine Policy* 71 (September 2016): 217–21.

Bender Shetler, Jan. "The Gendered Spaces of Historical Knowledge: Women's Knowledge and Extraordinary Women in the Serengeti District, Tanzania." *International Journal of African Historical Studies* 36, no. 2 (2003): 283–307.

Showers, Kate B., and Gwendolyn M. Malahleha. "Oral Evidence in Historical Environmental Impact Assessment: Soil Conservation in Lesotho in the 1930s and 1940s." *Journal of Southern African Studies* 18, no. 2 (June 1992): 276–96.

Semken, M., K. K. Miller, and K. James. "The Value of Oral History in Natural Resource Management: The Balcombe Estuary Reserve, Mount Martha, Victoria." *Australasian Journal of Environmental Management* 18, no. 2 (June 2011): 109–22.

Skeldon, Ronald. "Volcanic Ash, Hailstorms and Crops: Oral History from the Eastern Highlands of Papua New Guinea." *The Journal of the Polynesian Society* 86, no. 3 (September 1977): 403–9.

Skelton, Leona Jayne. "The Uncomfortable Path from Forestry to Tourism in Kielder, Northumberland: A Socially Dichotomous Village?" *Oral History* 42, no. 2 (2014): 81–93.

Sowards, Adam M. "'We're All Kinda Crazy': Smokejumpers and Western Forests." *Oral History Forum d'histoire orale 30*, Special Issue "Talking Green: Oral History and Environmental History" (2010): 1–22. https://www.oralhistoryforum.ca/index.php/ohf/article/download/389/459/0.

Stevens, Charles J. 'Taking Over What Belongs to God: The Historical Ecology of Tonga since European Contact." *Pacific Studies* 22, no. 3/4 (September/December 1999): 189–219.

Stevenson, Marc G. "Indigenous Knowledge in Environmental Assessment." *Arctic* 49, no. 3 (September 1996): 278–91.

Sutton, Lindsey, and Carla Restrepo. "Natural Hazards, Diverse Economy and Livelihoods in the Sierra de Las Minas, Guatemala." *Journal of Latin American Geography* 12, no. 3 (2013): 137–64.

Tauro, Alejandra, and Sergio Guevara Sada. "Historias Para Escuchar Y Diversidad QueCuidar." *Cuadernos de Biodiversidad* 29 (2009): 14–19. https://doi.org/10.14198/cdbio.2009.29.03.

Thurstan, Ruth H., Sarah M. Buckley, and John M. Pandolfi. "Oral Histories: Informing Natural Resource Management Using Perceptions of the Past." In *Perspectives on Oceans Past: A Handbook of Marine Environmental History*, edited by Bo Poulsen and Kathleen Schwerdtner Máñez, 155–73. Dordrecht, Netherlands: Springer, 2016.

Toosi, Fereshteh. "UP THE CREEK! A Site-Specific Oral History Project." *Oral History Forum d'histoire orale 30*, Special Issue "Talking

Green: Oral History and Environmental History" (2010): 1–3. https://www.oralhistoryforum.ca/index.php/ohf/article/download/391/461.

Twigg, Karen. "The Green Years: The Role of Abundant Water in Shaping Postwar Constructions of Rural Femininity." *Environment and History* 27, no. 2 (May 1, 2021): 277–301. https://doi.org/10.3197/096734021X16076828553539.

Watkins, Charles. "Methods and Approaches in the Study of Woodland History" In *Europe's Changing Woods and Forests: From Wildwood to Managed Landscapes*, edited by Keith Kirby and Charles Watkins, 18–31. Wellington, UK: Centre for Agriculture and Bioscience International, 2015.

Weaver, Sharon. "First Encounters: 1970s Back-to-the-Land, Cape Breton, NS and Denman, Hornby and Lasqueti Islands, BC." *Oral History Forum d'histoire orale* 30, Special Issue (2010): 1–30. https://www.oralhistoryforum.ca/index.php/ohf/article/download/387/458.

Williams, Brian, and Mark Riley. "The Challenge of Oral History to Environmental History." *Environment & History* 26, no. 2 (2018): 207–31. https://doi:10.3197/096734018X15254461646503.

Wolfe, Mikael. "'A Revolution Is a Force More Powerful Than Nature': Extreme Weather and the Cuban Revolution, 1959–64." *Environmental History* 25, no. 3 (July 2020): 469–91. https://doi.org/10.1093/envhis/emaa004.

Worster, Donald. "The Two Cultures Revisited: Environmental History and the Environmental Sciences." *Environment and History* 2, no. 1 (February 1996): 3–14.

THE OXFORD ORAL HISTORY SERIES

Rethinking Oral History and Tradition: An Indigenous Perspective *Nepia Mahuika*

Sisterhood and After: An Oral History of the UK Women's Liberation Movement, 1968-present *Margaretta Jolly*

Narrating South Asian Partition: Oral History, Literature, Cinema *Anindya Raychaudhuri*

Voices of Guinness: An Oral History of the Park Royal Brewery *Tim Strangleman*

Fly Until You Die: An Oral History of Hmong Pilots in the Vietnam War *Chia Youyee Vang*

Edward M. Kennedy: An Oral History *Barbara A. Perry*

The Land Speaks: New Voices at the Intersection of Oral and Environmental History *Edited by Debbie Lee and Kathryn Newfont*

The Voice of the Past: Oral History, Fourth Edition *Paul Thompson with Joanna Bornat*

When Sonia Met Boris: An Oral History of Jewish Life under Stalin *Anna Shternshis*

Inside the Clinton White House: An Oral History *Russell L. Riley*

Escape to Miami: An Oral History of the Cuban Rafter Crisis *Elizabeth Campisi*

Velvet Revolutions: An Oral History of Czech Society *Miroslav Vaněk and Pavel Mück*

Pioneers and Partisans: An Oral History of Nazi Genocide in Belorussia *Anika Walke*

Doing Oral History, Third Edition *Donald A. Ritchie*

A Guide to Oral History and the Law, Second Edition *John A. Neuenschwander*

Chinese Comfort Women: Testimonies from Imperial Japan's Sex Slaves *Peipei Qiu, with Su Zhiliang and Chen Lifei*

Listening on the Edge: Oral History in the Aftermath of Crisis *Edited by Mark Cave and Stephen M. Sloan*

Dedicated to God: An Oral History of Cloistered Nuns *Abbie Reese*

Lady Bird Johnson: An Oral History *Michael L. Gillette*

Bodies of Evidence: The Practice of Queer Oral History *Edited by Nan Alamilla Boyd and Horacio N. Roque Ramírez*

Soviet Baby Boomers: An Oral History of Russia's Cold War Generation *Donald J. Raleigh*

Freedom Flyers: The Tuskegee Airmen of World War II *J. Todd Moye*

Habits of Change: An Oral History of American Nuns *Carole Garibaldi Rogers*

They Say in Harlan County: An Oral History *Alessandro Portelli*

The Wonder of Their Voices: The 1946 Holocaust Interviews of David Boder *Alan Rosen*

Launching the War on Poverty: An Oral History, Second Edition *Michael L. Gillette*

The Firm: The Inside Story of the Stasi *Gary Bruce*

Singing Out: An Oral History of America's Folk Music Revivals *David K. Dunaway and Molly Beer*

Approaching an Auschwitz Survivor: Holocaust Testimony and Its Transformations *Edited by Jürgen Matthäus*

The Oxford Handbook of Oral History *Edited by Donald A. Ritchie*

Index

For the benefit of digital users, indexed terms that span two pages (e.g., 52–53) may, on occasion, appear on only one of those pages.

Note: Page numbers followed by f indicate a figure on the corresponding page.

Printed in the USA/Agawam, MA
December 5, 2022

802381.014